A History of Pesticides

A History of Pesticides

Graham A. Matthews

Emeritus Professor, Imperial College London

CABI is a trading name of CAB International

CABI
Nosworthy Way
Wallingford
Oxfordshire OX10 8DE
UK

Tel: +44 (0)1491 832111
Fax: +44 (0)1491 833508
E-mail: info@cabi.org
Website: www.cabi.org

CABI
745 Atlantic Avenue
8th Floor
Boston, MA 02111
USA

Tel: +1 (617)682-9015
E-mail: cabi-nao@cabi.org

A catalogue record for this book is available from the British Library, London.

Library of Congress Cataloging-in-Publication Data

Names: Matthews, G. A., author.
Title: A history of pesticides / Graham A. Matthews.
Description: Boston, MA : CABI, [2018] | Includes bibliographical references and index.
Identifiers: LCCN 2018034272 | ISBN 9781786394873 (hardback)
Subjects: LCSH: Pesticides--History.
Classification: LCC SB951 .S245 2018 | DDC 632/.95--dc23
LC record available at https://lccn.loc.gov/2018034272

ISBN: 978 1 78639 487 3 (Hardback)
978 1 78639 489 7 (PDF)
978 1 78639 490 3 (ePub)

Commissioning editor: Ward Cooper
Editorial assistant: Alexandra Lainsbury/Tabitha Jay
Production editor: Tim Kapp

Typeset by SPi, Pondicherry, India
Printed and bound in Great Britain by Severn, Gloucester

Contents

Preface

Having specialized in pesticide application technology used in crop protection and vector control for six decades, this book is a look back at the history of pesticides, going far beyond the period during which I have worked. I have been helped by previous books on the subject, starting with E.G. Lodeman's *The Spraying of Plants*. In a preface to that book, R.T. Galloway, the Chief of the Division of Vegetable Pathology, USDA, remarked on the 'rapid advance made in combating the insects and fungi that attack our cultivated plants'. That was in the early days of Bordeaux mixture. Advances did not happen quickly until 60 years later with DDT and 2,4-D, coming as World War II was being fought. The rapid uptake of DDT to control mosquitoes and insect pests on crops was described by A.W.A. Brown in his book *Insect Control by Chemicals*. Other early books containing useful information were Samuel Potts' *Concentrated Spray Equipment, Mixtures and Application Methods* (1958) and Graham Rose's *Crop Protection* (1963).

My own career began in 1958 when I was recruited to join an entomological team tasked with developing control of insect pests in cotton for the small-scale farmers in the Federation of Rhodesia and Nyasaland. The intention was to include a chemist, and as I had done some chemistry at university I became the third entomologist on the team, specifically 'to help the farmers with DDT as had been done in Texas'. This was before Rachel Carson's book *Silent Spring*. Before that book was published, we had initiated trials on farms comparing the existing unsprayed cotton with a programme using carbaryl for red bollworm control and DDT for the so-called American bollworm, and shown that it was possible to double or triple yields depending on the rainfall. Later, higher yields were possible with irrigation. Rachel Carson was stimulated to write her book by the large number of birds dying both in the USA and elsewhere, which was shown to be due to them eating seeds treated with an organochlorine insecticide, which protected young plants from soil pests. The thinning of their egg shells (Radcliffe, 1967) was another factor in the decline in bird populations. Her book led eventually to the Stockholm Convention. She says:

> It is not my contention that chemical insecticides must never be used. I do contend that we have put poisonous and biologically potent chemicals indiscriminately into the hands of persons largely or wholly ignorant of their potentials for harm...If we are going to live so intimately with these chemicals...eating and drinking them...we had better know something about their nature and power.
>
> (Carson, 1962)

Over the last 60 years, pesticides have undoubtedly played a key role in protecting farmers' crops, and will continue to do so, albeit increasingly as a last resort in integrated pest management programmes, while we enter a new era in which new technologies will play an increasingly important role. Already the development of genetically modified crops has demonstrated the ability to incorporate an insecticidal toxin into plants to minimize the need to spray insecticides, and to develop herbicide-tolerant crops to facilitate a simpler weed management programme. Our increasing knowledge of the genome of crops and pests will inevitably lead to a new era in crop protection, together with new varieties that provide drought tolerance and better nutrition, as well as greater resistance to pests. The hope is that, when using pesticides, we progress beyond the present systems of application that have evolved from the 19th-century spray nozzles and ensure that applications are more targeted.

Throughout this book, the common names of pesticides have been used with a few trade names sometimes mentioned. There are so many different trade names for a single active ingredient that it would be difficult to include all of them. The use of any of the pesticides mentioned depends on whether the active ingredient and local trade name has been approved and registered in a particular country.

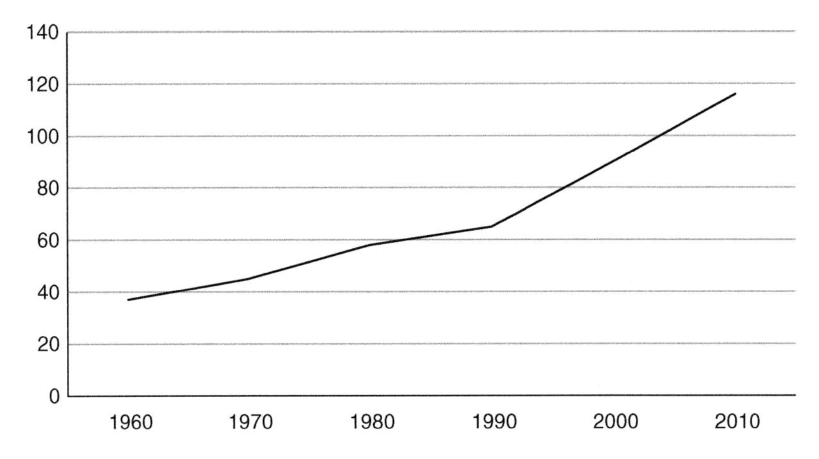

Fig. 1. An index showing the increase in global crop production from 1960 [2004 = 100], part of which is due to increased protection of crops using pesticides during the Green Revolution.

I have not given the detailed chemical structure of pesticides as this information is easily obtained from the internet. The internet also provides access to detailed information about pesticides such as the Pesticide Action Network (http://www.pesticideinfo.org) and the pesticide database at Hertford University (https://sitem.herts.ac.uk/aeru/ppdb/en).

Many of the photographs are from my own collection, but some have been copied from the internet, where use of the photograph is licensed under the Creative Commons Attribution 2.5 Generic License (https://creativecommons.org/licenses/by/2.5/).

References

Brown, A.W.A. (1951) *Insect Control by Chemicals.* John Wiley, New York; Chapman & Hall, London.

Carson, R. (1962) *Silent Spring.* Houghton Mifflin Co., Cambridge, Massachusetts, USA.

Epstein, L. (2014) Fifty years since *Silent Spring. Annual Review of Phytopathology* 52, 377–402.

Fletcher, W.W. (1978) *The Pest War.* Wiley, Chichester, UK.

Lodeman, E.G. (1896) *The Spraying of Plants.* Macmillan, London.

Potts, S.F. (1958) *Concentrated Spray Equipment, Mixtures and Application Methods.* Dorland Books, Caldwell, New Jersey, USA.

Radcliffe, D.A. (1967) Decrease in eggshell weight in certain birds of prey. *Nature* 215, 208–210.

Rose, G. (1963) *Crop Protection.* Leonard Hill, London.

Acknowledgements

Having been involved with the application of pesticides for 60 years, it seemed an opportune moment to reflect on their use. Inevitably, the methods of application form an important part of the history of their use, but the chemicals that have been developed and marketed, as well as the more recent attention on biopesticides over the period of my career, form the basis of this book. I have been helped enormously by the comments and guidance of colleagues on individual chapters. I am indebted to Terry Wiles for his assistance with three of the first four chapters, especially his interest in herbicides and no-till agriculture. Simon Archer has been of considerable help with all the sections dealing with fungicides. John Tunstall, with whom I began my career in the Federation of Rhodesia and Nyasaland, has contributed significantly to several chapters. Denis Wright has helped me with the chapters on insecticides post-1950 and pesticide resistance, while Helmut van Emden contributed to the IPM chapter. Keith Walters, Roy Bateman and Len Copping commented on Chapters 6, 9 and 10, respectively.

Many of the illustrations from the 19th century are from Lodeman's book *The Spraying of Plants*. I thank the Agricultural Experiment Station, Wooster, Ohio, for permission to record the first use of an aircraft applying pesticide dusts in 1921. Some of the illustrations are copied from Google Images, with permission sought where the original source was evident. Photographs showing preparation of tobacco seed beds were kindly supplied by Kutsaga Tobacco Research Station in Zimbabwe. Illustrations in Chapter 10 were scanned from *The History of Pest Control*, kindly lent to me by Bruce Knight. I thank Sensat Company, London, and Yamaha-Motor for permission to use photographs of drones. The assistance of Rentokil, Kutsaga Research Station and Bioline AgroSciences for specific illustrations is much appreciated. I thank the manufacturers of application equipment, including Househam, Micron Sprayers Ltd and others who supplied photographs for my previous books. Many illustrations are from my own collection. Other illustrations from journals were permitted under open access or by the publisher.

Special thanks are due to my wife, Moira, who has been supporting my work on pesticide use for over 50 years.

References

Fisons (1976) *The History of Pest Control.* Fisons Ltd.
Lodeman, E.G. (1896) *The Spraying of Plants.* Macmillan, London.

Prologue – Before Pesticides

Some examples of the enormous impact of pests, diseases and weeds on humans reveal enormous loss of life, failure of crops and drudgery in efforts to control them that can now be averted or at least minimized by the scientific development and sensible use of pesticides.

Even in biblical times the devastating impact of pests was recognized: 'What the palmerworm left, the swarming locust has eaten. What the swarming locust left, the hopping locust has eaten, and what the hopping locust has left, the destroying locust has eaten' (Joel 1:4). Depending on the translation, different locusts or caterpillars are referred to as eating all the vegetation. According to Howard (1931), locusts were the cause of famine in Algeria in 1866 when 5% of the population died. According to Dr Uvarov, globally, locusts caused an estimated loss of £15 million annually before World War II (Ordish, 1952).

Perhaps one of the most devastating diseases was bubonic plague, caused by a bacterium, *Yersinia pestis*, which killed an estimated 50 million people in Europe between 1346 and 1353 and continued in some areas until 1654. The disease, known as the Black Death, was spread to people by fleas that had fed on infected rats living close to humans. When bitten by infected fleas, the bacterium develops and forms a painful swelling, often in the groin or on the thigh, armpit or neck. Eighty per cent of those infected died, usually within three weeks, and during the warmer summer months between July and late September. In the UK, a third of the population died from bubonic plague, with a catastrophic impact on trade and the economy, especially in rural areas. Outbreaks of plague did not occur during the winter months, as low temperatures reduced the activity of the fleas. It was thought that the disease originated in an area close to the Caspian Sea. It was spread through the Eurasian steppes by rats, gerbils and possibly camels, through the arid and semi-arid landscape. Later, rats on board ships crossing the Black Sea, and then the Mediterranean, gradually spread the disease north to other areas including Russia. Finland and Iceland were the two areas that avoided the plague, presumably because temperatures did not favour the fleas. In those days there was no insecticide to control the fleas, nor rodenticide to kill the rats. Plague is still present and has been reported in Madagascar and several other countries, but at least the disease can now be checked with antibiotics.

The lack of a fungicide resulted in the devastation of the coffee industry in Ceylon (Sri Lanka) at the end of the 19th century. Growing coffee in Ceylon had been started in 1740 by the Dutch, but expanded after the British took over the country, encouraged by demand for coffee in Europe. Large areas were deforested to allow for the increase in coffee plantations. The country became one of the major coffee-producing nations in the world, with a peak in production in 1870; over 100,000 hectares were cultivated. It was then that the fungal disease Coffee leaf rust (*Hemileia vastatrix*) arrived, allegedly as a result of a British military expedition from the Sudan, which passed through Abyssinia (now Ethiopia), the ancestral homeland of both *Coffea arabica* and its leaf rust, resulting in such a severe decline in production that growers switched to the production of tea. Although there is still some coffee grown in Sri Lanka, its production was ranked only 43rd in the world in 2014. Tea (*Camellia sinensis*) was not susceptible to the disease, so growers were able to expand production, making Sri Lanka a leading worldwide exporter of tea.

Another country, Ireland, also suffered from the lack of a fungicide, when the disease potato blight (caused by *Phytophthora infestans*) is thought to have come from the USA by sea after spreading from Mexico. The potato crop had failed periodically due to disease or frost prior to 1840, but the devastating impact of blight led to the Great Famine with mass starvation, the death of a million people and the subsequent migration of another million between 1845 and 1852, resulting in a 20–25% drop in population. The severe impact of blight in Ireland, compared with other parts of Europe, was probably due to the potato being an essential part of the Irish diet (cereals being difficult to grow in the wet climate) and a lack of genetic variability among the potato plants.

In the absence of suitable insecticides, the arrival of the boll weevil (*Anthonomus grandis*) from Mexico around 1892 had a major impact on cotton production in the southern states of the USA. From Texas, boll weevils spread northwards very rapidly, reaching Arkansas and Mississippi in 1907; and by 1922, 85% of the cotton growing area was affected. Damage to the Texan cotton crop in 1903 was conservatively estimated at $15 million. The only area that expanded production, partly due to the absence of boll weevil, was in the west of the USA. Once insecticides became available they were used to minimize the impact of the weevils. Initially, calcium arsenate dusts were applied from around 1923 until the 1950s, when low-volume sprays were applied, but the huge costs involved led to a major programme aimed at eradicating the pest from the USA.

Prior to the development of herbicides, one of the main tasks on farms was removing weeds from fields, which resulted in a huge demand for human labour. Thus, in the USA around 1850, 65% of the population lived on farms to weed crops (Gianessi and Reigner, 2007), despite leaving fields fallow and rotating crops to new land in an attempt to reduce weeds. The development of equipment to cultivate fields using animals, and then tractors, increased the use of mechanical control of weeds, until the rapid adoption of herbicides in the 1950s replaced the millions of workers who hoed weeds by hand or used mechanical tillage. Herbicides were cheaper and more effective than hand weeding and cultivation, thus reducing production costs and increasing yields. Even today there are many areas, particularly poorer areas of the Tropics, where areas of crops are abandoned if there is insufficient labour for hand weeding during the crucial first few weeks of crop growth.

Early attempts to use a pesticide

Lodeman (1896) recorded some of the earliest instances of plants being protected from diseases and insect pests. In 1629, John Parkinson recommended using vinegar to prevent canker on trees. In *Paradisi in Sole Paradisus Terrestris* he states that 'Canker is a shrewd disease... and must be looked into in time before it hath run too farre: most men doe wholly cut away as much as is fretted with the canker and then dresse it or wet it with vinegar... '. Reference was also made to the use of a quart of common salt in 2 gallons of water, and when all the salt had dissolved the brine was used to wash scale insects on trees. Around the same time, Austen (1653), in *A Treatise of Fruit Trees*, recommended washing cankered branches with cow urine and, more helpfully as a source of potassium, dressing the surrounding soil with wood ashes.

Early attempts to develop remedies were often for use against human and animal pests. In the 17th century, a Mr Tiffin established a company in Hatton Garden, London, and contracted to keep beds free from bedbugs for the sum of 3 shillings per year. The company had a royal warrant and a policy limiting it to only 100 customers. Interestingly, in 2005, a new company, Bed Bugs Ltd, was set up to emulate the original Tiffin & Son's service in London.

In 1711, it was suggested that an insect, Cantharides (*Lytta vesicatoria*), or Spanish fly, an emerald green beetle on trees such as ash, could be destroyed by using a pump to wet them with water that had been boiled with 'some rue'. The common rue or herb-of-grace (*Ruta graveolens*), a native of the Balkan Peninsula, is grown as an ornamental and as a herb. Perhaps its very disagreeable odour and sharp, bitter taste were thought to make it a good insecticide.

In 1763, a method of application using a small tin syringe having a nose pierced with about 1000 holes was described for applying a handful of finely powdered bad tobacco mixed with 2 l of water and in which lime was then slaked. It was recommended that this treatment was repeated after 4–5 days to kill plant lice (Goeze, 1787) and is possibly the first use of nicotine as an insecticide.

Many other recipes were tried. Forsyth (1802) had been trying a mixture of cow dung and lime; some were using a soap or urine, but he recommended half a peck[1] of unslaked lime in 32 gallons of water allowed to stand for 3–4 days before being applied with a syringe to control aphids. Whale oil soap was another remedy, and sulfur was used against some diseases. In 1843, a Mr William Cooper marketed a product with arsenic and sulfur to cure sheep scab. He later marketed Cooper's Wheat Dressing, a product containing arsenic and soda ash, sold at 6d a packet to treat six bushels to control smut, a disease noted by Jethro Tull when he developed a drill to sow three rows of wheat and turnip seed in drills at a time (Tull, 1743). By 1870 he was selling sufficient amounts to treat about 100,000 acres per year. Much later, the company he established became Cooper, McDougall & Robertson Ltd, which merged with a subsidiary of ICI – Plant Protection Ltd – in 1937.

Phylloxera (*Daktulosphaira vitifoliae*) was introduced into France on vines brought in from the USA in the late 1860s. The damage caused by these pale yellow sucking insects, similar to aphids, feeding on the roots and leaves, devastated vineyards, so in France they attempted to graft American root stock to their own vines in order to produce a more resistant strain of grape. Around this time, in 1878, downy mildew (caused by *Plasmopara viticola*) on grapevines was first noted in France on some of the American grape seedlings. However, some owners of vineyards were also suffering losses caused by children and travellers taking grapes alongside the highways. To discourage the theft they sprinkled a mixture of milk of lime and copper sulfate, using a brush (Fig. 2) to colour the vines blue and make the ripening grapes appear to be poisoned. The protective effect of this against downy mildew was soon observed, notably by Millardet (Fig. 3), a chemistry professor at Bordeaux University, and led to the development of Bordeaux mixture. An early recipe was to dissolve 8 kg of commercial sulfate of copper in 100 l of water, and in a separate vessel make a milk of lime by slaking 15 kg of quicklime in 30 l of water. This was added to the copper sulfate solution to form a bluish precipitate that was stirred well. Some, carried in a pail, was sprinkled on the vines using a small broom. This proved to be very successful in 1885 when the downy mildew was very intense and defoliated untreated vines. Various formulas were tried, one adding glue, which was apparently beneficial.

Another sulfate that really started to be used after 1900 was ferrous sulfate, which is still used today to control moss in lawns and in turf management. It may also be sold mixed with fertilizer to encourage strong root development of grass and tillering to cover where moss has been present. Some 50 years after ferrous sulfate was used on lawns, it is also available mixed with certain herbicides, such as dichlorprop-P and MCPA to control weeds in lawns.

Fig. 2. Brush used to sprinkle copper sulfate on vines in France to deter children stealing grapes.

Fig. 3. Professor Millardet, Bordeaux University. (Photo from Lodeman, 1896, used with permission)

In the early days of using Bordeaux mixture, there was considerable interest in development of spraying equipment. In the USA, the first ‘knapsack sprayer’ had two hoses attached to the bottom of the tank so that two rows were treated as the liquid was gravity fed to the ‘sprinkler’ on the end of the hose (Fig. 4).

This had been developed to apply Paris green on potatoes that were infested with the Colorado beetle, an alien pest from Mexico, which had become so serious that spraying the crop was widespread by 1875. In France, around 1885, a knapsack sprayer using a pump was designed, and by 1890 some were imported into the USA. The Vermorel ‘Éclair’ had a rubber disc to form a diaphragm pump (Fig. 5), while the ‘Vigourex’ had a piston pump.

The Japy and Albrand were competitors, the latter having an air pump and separate reservoir, thus being the forerunner of the compression sprayer. Soon, a knapsack sprayer, the ‘Galloway’, was designed and manufactured in the USA. At the same time, various nozzles were designed to provide a straight jet or a cone, or variable cone, of spray. Larger, wheeled equipment was soon developed, but it relied on manual pumping of the spray. However, one design called a potato sprayer was fitted with revolving horizontal brushes fed by gravity from the spray tank (Figs 6–9).

According to Lodeman (1896) the best spray was said to be one that nearly resembles a fog, but it was noted that:

Fig. 4. Treating potatoes with Paris green in the USA to control Colorado beetles, mid-1800s.

Fig. 5. The Vermerol Éclair Sprayer to apply Bordeaux mixture.

Fig. 6. Barrel sprayer. (From Lodeman, 1896, used with permission.)

Fig. 7. Wheeled sprayer. (From Lodeman, 1896, used with permission.)

Fig. 8. Horsedrawn sprayer, *c.*1900. (From Lodeman, 1896, used with permission.)

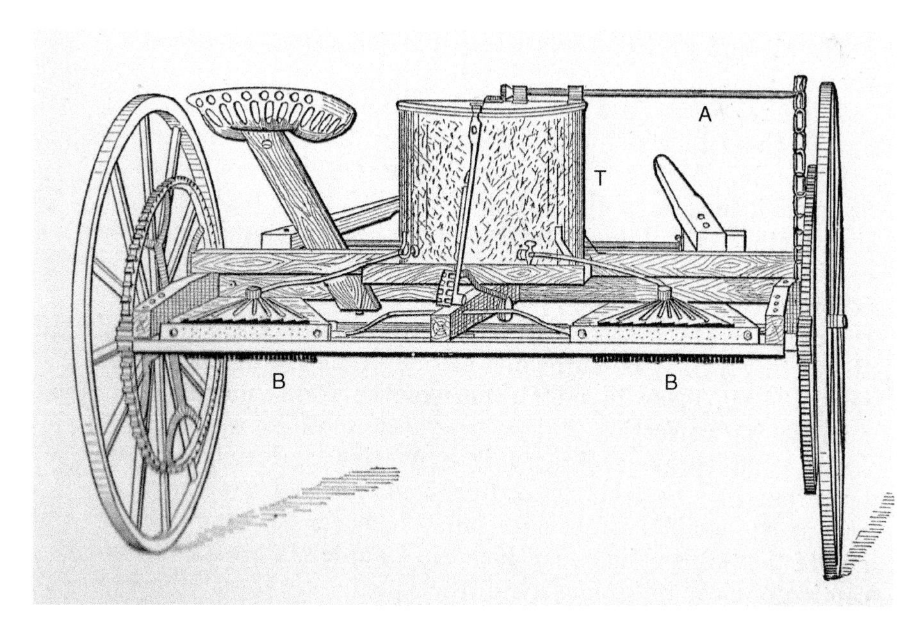

Fig. 9. Potato sprayer with rotary brush fed by gravity. (From Lodeman, 1896, used with permission.)

> ...the finer the spray, the less liquid is thrown, and the smaller the area treated. Whenever the wind blows, a fog-like spray will go wherever the wind carries it, and not where the operator directs it. Sometimes this will be an advantage....Yet when the wind will come from the wrong direction, much of the material is blown where it is not wanted.

This would appear to be an early recognition of spray drift.

Note

[1] A peck may be used for either liquid or dry measure and is equal to 8 imperial quarts (2 imperial gallons) or a quarter imperial bushel, or 554.84 in^3 (9.092 l).

References

Austen, R.A. (1653) *A Treatise of Fruit Trees.* Thomas Robinson, London.
Forsyth, W. (1802) *A Treatise on Culture and Management of Fruit Trees.* US edition edited by William Cobbett. D. & S. Whiting, Albany, New York.
Gianessi, L.P. and Reigner, N.P. (2007) The value of herbicides in US crop production. *Weed Technology* 21, 559–566.
Goeze, J.A.E. (1787) *Geschichte einiger, den Menschen, Thieren, Oekonomie und Gärtneren schädlichen Insekten.* Weidmanns Erben und Reich, Leipzig, Germany.
Howard, L.O. (1931) *The Insect Menace.* Appleton & Co., London.
Lodeman, E.G. (1896) *The Spraying of Plants.* Macmillan, London.
Ordish, G. (1952) *Untaken Harvest.* Constable, London.
Parkinson, J. (1629) *Paradisi in Sole Paradisus Terrestris.* Methuen, London.
Tull, J. (1743) Horse-hoeing husbandry. In: Ordish, G. (ed.) *Untaken Harvest.* Constable, London.

1 Pesticides in the Early Part of the 20th Century

After the many different attempts to develop pesticides during the 19th century, efforts over the early decades of the 20th century concentrated on two main areas – the use of extracts from plants, notably pyrethrum and tobacco, and certain inorganic chemicals, mostly containing arsenic, sulfur or copper. Then, from the 1940s onwards, chemists started to develop organochlorine and organophosphate insecticides as well as new herbicides and fungicides. A brief overview of the pesticides used from 1900–1960 is given in this chapter.

Botanical Insecticides

In grasslands and forests, plants are able to survive, as they contain chemicals that enable them to combat attacks from insects and diseases. The main plants that man has selected over the centuries as food plants generally have very low levels of toxins. The earliest insecticides were essentially dried leaves of some plants, and, ultimately, modern science has played an important role in identifying these botanical insecticides and subsequently developing similar chemicals that are more effective, photostable and economical to market to farmers. One important food crop, Cassava (*Manihot esculenta*), is very poisonous unless the roots are well cooked. Farmers often prefer the bitter varieties because they deter pests.

Pyrethrins

Pyrethrum was known as far back as 400 BC in Persia (now Iran) and it was thought to have been used in stores, but interest in pyrethrum in Europe increased early in the 19th century, apparently due to an Armenian who learnt about the powder. In Europe it was initially referred to as 'Dalmatian powder' obtained from the flower heads of *Pyrethrum cinerariaefolium* (now called *Chrysanthemum cinerariaefolium*), grown in the Balkans. An interesting story is that a German woman in Dubrovnik, Dalmatia, picked the flowers to have in her house, and having discarded the withered flowers in the corner of a room she noticed later that the plants were surrounded by dead insects, and apparently associated this with insecticidal properties of the plants. By 1850 the powder was used to kill insects in houses in France. Bales of dried flowers and seeds were exported to the USA where the powder was used in dwellings and glasshouses. The main source of disruption in supply of pyrethrum, caused by World War I, was Japan, where the crop had been grown since 1886; but after World War II, Kenya took over the main production (Fig. 1.1). In 1917, the US Navy mixed a pyrethrum extract with kerosene to produce a space spray to control house flies and mosquitoes (Glynne-Jones, 2001). Globally, there are over 2000 registered products containing pyrethrins, used mostly in homes and for controlling mosquitoes, for example in mosquito coils and domestic sprayers such as the Flit gun, used prior to aerosols.

Studies on pyrethrum around 1910–1916 by Staudinger and Ruzucka (1924) separated and partially identified the two primary active principles of pyrethrum – Pyrethrin I and Pyrethrin II. This led to considerable research on these actives (Gnadinger, 1936). Pyrethrin I was considered to be more toxic than Pyrethrin II, but the latter was far superior in causing 'knock-down' of house flies (Sullivan, 1938). Studies by Tattersfield (1931) and others continued, as its use had proved to be very effective indoors, enhanced later by the development of piperonyl butoxide (PBO) as

Fig. 1.1. Chrysanthemum flowers for extraction of pyrethrum.

a synergist in the late 1930s and early 1940s, but it was not photostable, so research continued until the first photostable permethrin was developed (Elliott *et al.*, 1973), followed by other pyrethroids, discussed later.

Piperonyl butoxide was developed in the late 1930s and early 1940s to enhance the performance of the naturally derived insecticide pyrethrum. As a synergist, it inhibited the natural defence mechanisms of the insect, especially the mixed-function oxidase system (MFOs) also known as the cytochrome P-450 system. It was not considered necessary to use PBO when the synthetic pyrethroids were developed, but when *Helicoverpa armigera*, the cotton bollworm in Australia, became resistant to pyrethroids, following their extensive use, studies on using PBO revealed that esterase inhibition did not occur until 3–4 hours after PBO had been applied, suggesting a need for a pre-treatment prior to the pyrethroid spray (Young *et al.*, 2005). More recently, PBO has been added to bed nets treated with pyrethroids to increase the mortality of the mosquitoes resistant to pyrethroids.

Rotenone

Rotenone is another botanical insecticide, known for centuries. The Chinese had extracted the insecticide from the roots of a vine growing wild in Asia, known as derris (*Derris elliptica*), but it is also found in devil's shoe string *(Cracca virginiana)* (Roark, 1933) and other plants – *Tephrosia*, *Millettia*, *Mundulea* and *Pachyrhizus* (Brown, 1951). It had also been used as a poison dip for arrows in Borneo, but was best known as a fish poison. In 1902, a Japanese chemist isolated the most potent insecticidal substance in derris and called it rotenone. The neurotoxin had been regarded as harmless to human beings but 15 times more toxic to aphids than nicotine. Derris, supplied as a liquid or dust, was generally available for gardeners and 'organic' vegetable growers in the UK until October 2009 when it ceased following an EU Directive. Rotenone has been used for the management of invasive fish species, but there is concern, as this also affects non-targeted organisms including amphibians and macro-invertebrates (Dalu *et al.*, 2015).

Nicotine

The alkaloid nicotine is found in many solanaceous plants, notably in the leaves of *Nicotiana rustica*, in amounts of 2–14%, in the tobacco plant *Nicotiana tabacum*, the Australian pituri (*Duboisia hopwoodii*) and common milkweed (*Asclepias syriaca*as). Its use as an insecticide started with tobacco leaves. As Lodeman (1896) mentioned, two handfuls of Virginia tobacco mixed with a handful of wormwood and a handful of rue in two pailfuls of water, boiled for half an hour and then strained, was ready to be sprayed. Tobacco alone was good, but not as good as the mixture. Later it was usually marketed as nicotine sulfate, which is non-volatile but becomes so in proportion as it is changed to nicotine by the addition of an alkali to neutralize the combining acid. Its toxicity was mainly due to the 'fumigation' effect (de Ong, 1924). In 1880, a Mr G.H. Richards set up a company to market a standardized product, XL Nicotine, suitable for gardeners to use as it was successful in controlling sucking pests including mealybugs, woolly aphids and certain scale insects with a waxy cuticle, due to the penetration of the vapour. It was more effective if the ambient temperature exceeded 16°C.

As nicotine, like other botanical extracts, is not persistent, recent interest has been in the use of nicotinoid insecticides (Ujváry, 1999), generally referred to as the neonicotinoids, which are discussed later (see Chapter 3).

Ryania

The botanical insecticide ryania is the ground stem wood and roots of the salicaceous plant *Ryania speciosa*, a plant originally recorded as found in Trinidad (Brown, 1951). The insecticidal activity of ryania extract was attributed to ryanodine but later shown to be due to the combination of ryanodine and the equipotent and more abundant 9,21-dehydroryanodine (Jefferies *et al.*, 1992). It was very effective in controlling European corn borer and the sugar cane borer. As with other botanical insecticides, there are now modern synthetic ryanoids, which include chlorantraniliprole, cyantraniliprole and flubendiamide.

Inorganic Chemicals

Arsenicals

The use of arsenical poisons for crop protection was initiated by the arrival of major insect pests in the USA, for example the potato beetle, which arrived from Mexico according to some reports; but Lodeman (1896) refers to it as a native of the Rocky Mountains, which spread eastwards when growing potatoes had spread west into territory occupied by the beetle. It is now referred to as the Colorado beetle (*Leptinotarsa decemlineata*). It was such a vigorous feeder that farmers had to apply an insecticide. Paris green appeared around 1860 and became a standard insecticide, its use extending to other crops. It was also used to kill mosquito larvae. The name Paris green originates from its use as a rodenticide to kill rats in the sewers of Paris, competing with another arsenical, London purple, a less expensive by-product of the dye industry, which was exported in considerable quantities to the USA from 1878 by Messrs Hemingway & Co., London (Ordish, 1952). At that time, Paris green, referred to as emerald green, was also a popular pigment used in artists' paints.

According to Lodeman (1896), Paris green, a copper acetoarsenite, could be prepared by boiling a solution of white arsenic in one vessel and a similar one of acetate of copper (verdigris) in another. These two boiling solutions were combined and Paris green precipitated. The fine crystalline powder with a clear green colour was practically insoluble in water.

In the USA, dusting cotton with calcium arsenate to control the boll weevil (*Anthonomus grandis*) and the cotton leafworm (*Alabama argillacea*) began in the 1920s and was soon carried out in all the cotton growing states. Dusts were used instead of sprays, as arsenicals used were insoluble in water. They wanted the deposit on the foliage, so that it was ingested by insects and phytotoxicity was minimized (Brown, 1951). Nicotine dust was added to control the cotton aphid (*Aphis gossypii*) in 1926, and by 1930 the technique was used by the Russians in central Asia, although they applied calcium arsenite and then sulfur to control mites (*Tetranychus telarius*).

The quantities needed to control gypsy moth (*Lymantria dispar*) with Paris green proved very phytotoxic, so a change was made to lead arsenate, which was less soluble. Lead arsenate had been prepared as an insecticide much earlier, in 1892, for use against gypsy moth, but its use in forests began with aerial spraying, which commenced in Massachusetts in 1926. It was also aerially applied in the UK, as a dust, in 1922 on an orchard near Sevenoaks. Lead arsenate (LA) was the most extensively used of the arsenical insecticides but, for some pests, was replaced by the less expensive calcium arsenate, until DDT became widely available in 1948.

When the malaria vector *Anopheles gambiae* spread to north-east Brazil in the 1930s, Paris green was widely used as a larvicide (Killeen *et al*., 2002). Similarly, in Palestine, considerable efforts at improving drainage were supplemented by applying Paris green, the larvicide of choice from 1926 until 1948 (Kitron and Spielman, 1989), when DDT was used. In the Tennessee valley, in 1938, 95,000 acres of Wheeler Reservoir were dusted from the air with Paris green, and 4800 miles were oiled from the surface. Paris green was heavily sprayed by plane in Italy, Sardinia and Corsica during 1944, and in Italy in 1945, to control malaria.

Sulfur

Sulfur has been known to be effective against diseases such as rust on wheat since the Greek poet Homer described the benefits of 'pest-averting sulphur' 3000 years ago. Farmers continue to use sulfur dust to control plant diseases such as powdery mildew. In Tanzania, sulfur dust was recommended to treat cashew nut, a major cash crop, to control the powdery mildew disease caused by *Oidium anacardii* Noack. The standard recommendation in the 1980s was to apply 1.25 kg of sulfur dust per tree per season, so that with a tree spacing of 12×12 m, 90 kg of dust was applied per hectare spread over 4–5 applications using a motorized duster. To minimize the possible impact of acidification of the soil, Smith and Cooper (1997) proposed that the current dusting strategy could be improved by treating only a portion of the trees at each dusting round and spreading the applications over the mildew control season.

The effectiveness of sulfur as a fungicide could be increased by adding lime, which helped the sulfur penetrate plant tissues, as noted in 1851 by Monsieur Grison at Versailles, where he needed a product that was better than sulfur dust to combat the vine powdery mildew in his greenhouses. The mixture was prepared by heating an aqueous suspension of one part lime (calcium hydroxide) with two parts by weight of elemental sulfur (S). The mixture produced contained mostly calcium polysulfides with some calcium thiosulfate and some unchanged elemental sulfur. Many apple growers applied lime sulfur to control apple scab, often as a 'winter' spray before the buds opened.

Although yellow sulfur had been a proven organic treatment against powdery mildew on ornamentals, as well as on fruit and vegetables, its use was banned within the EU and other countries in 2011. It could still be used in soil as an acidifier or nutrient treatment. Sulfur has also been used to control insect pests, sometimes mixed with DDT or rotenone, and to control ticks on cattle and mites, for example on cotton; but when used in apple orchards, it had a detrimental impact on some important predators of codling moth and was also mildly phytotoxic on some crops.

Other inorganic chemicals

A number of other chemicals were used. Cockroaches, e.g. *Periplaneta* spp., were controlled using a bait containing less than 5% boric acid or as a dust. The bait has to be ingested to be effective. It is very toxic to young children and pets so great care is needed in using it in cracks and crevices under sinks and other sites favoured by cockroaches. Thallium acetate or thallous sulfate were used in baits to control ants. Some soil pests, such as cabbage root fly larvae (*Delia radicum*) were controlled with mercurous chloride (Calomel). Sodium selenite was applied as a systemic insecticide and acaricide. Generally, none of these compounds is now recommended.

Organic Chemicals

Some farmers and consumers have shown a preference for organically grown food, shunning the use of pesticides; but the new generation of pesticides, post-war, utilized organic chemistry rather than continuing with lead arsenate and other inorganic poisons. The Soil Association, formed in 1946, has been a keen advocate of organic farming and avoiding use of modern pesticides, especially after Rachel Carson's book *Silent Spring* was published in 1962.

DDT

The organochlorine insecticide DDT (dichlorodiphenyltrichloroethane) had been synthesized as early as 1874, but its insecticidal activity was not recognized until 1939. It was developed by Paul Hermann Müller at Geigy in Switzerland because 'the only available insecticides were either expensive natural products or synthetics ineffective against insects; the only compounds that were both effective and inexpensive were arsenic compounds, which were just as poisonous to human beings and other mammals'. He sought to 'synthesize the ideal contact insecticide – one which would have a quick and powerful toxic effect upon the largest possible number of insect species while causing little or no harm to plants and warm-blooded animals'. He also wanted a chemical that was stable and inexpensive (Roberts *et al.*, 2010). After four years of searching and trying over 300 chemicals, he found a chemical that 'when a fly was placed in a cage laced with it, the fly died a short while later'.

Geigy patented DDT in 1940 and marketed dust formulations Gesarol and Neocid. Some was distributed to the UK and used by the British Ministry of Supply in 1943 and by the US army during World War II. A considerable effort was made to examine DDT to determine its mode of action (Wigglesworth, 1955), its toxicology (Hayes, 1959) and its safety, it being used to control vectors of diseases on humans (Simmonds, 1959). DDT was tested as a residual insecticide against adult vector mosquitoes, and in Italy it was applied to the interior surfaces of all habitations and outbuildings of a community to test its effect on *Anopheles* vectors and malaria incidence. An early discovery was the initial nervous response of mosquitoes to DDT, which was to fly away before they had picked up a lethal dose (Kennedy, 1947). Thus it exhibited excitant and repellent properties, which resulted in many mosquitoes leaving sprayed houses without biting. DDT was initially used by the military to control malaria, typhus, body lice and bubonic plague, and in 1944, 3 million people in Naples were treated with DDT dust – approximately 22 g/person – to check an outbreak of typhus (Soper *et al.*, 1947) (Figs 1.2–1.4). The impressive achievement of defeating the spread of typhus led to Dr Müller being awarded the Nobel Prize for Medicine in 1948, 'for his discovery of the high efficiency of DDT as a contact poison against several arthropods'.

In 1945, Missiroli (1948) planned to eradicate malaria from Italy, and, the following year, started spraying a 5% solution of DDT in kerosene at 2 g a.i./m^2. By 1948, 4 million people had been protected by using 335 t of DDT at 1.5 g a.i./m^2 in 2.85 million premises (Pampana, 1951). From 75,000 malaria cases in Sardinia in 1946, use of DDT had reduced the number of cases to nine by 1951. Soon DDT was being recommended for controlling malaria worldwide by the WHO. In 1945, malaria infected an estimated 75 million people and killed 800,000 in India, but by the early sixties the number of cases had dropped to about 50,000. Similarly, in Ceylon (now Sri Lanka), the introduction of DDT reduced the number of malaria cases from 2.8 million in 1948 to fewer than 30 in 1964. Berry (1990) pointed out that over a ten-year period, the WHO programme of spraying DDT used around 400,000 tonnes without evidence of toxicity to operators and with a calculated saving of 15 million lives. The selection of mosquitoes resistant to DDT resulted in the WHO withdrawing the global

Fig. 1.2. DDT dust being applied to troops in World War II. (Photo courtesy of H.D. Hudson Manufacturing Co.)

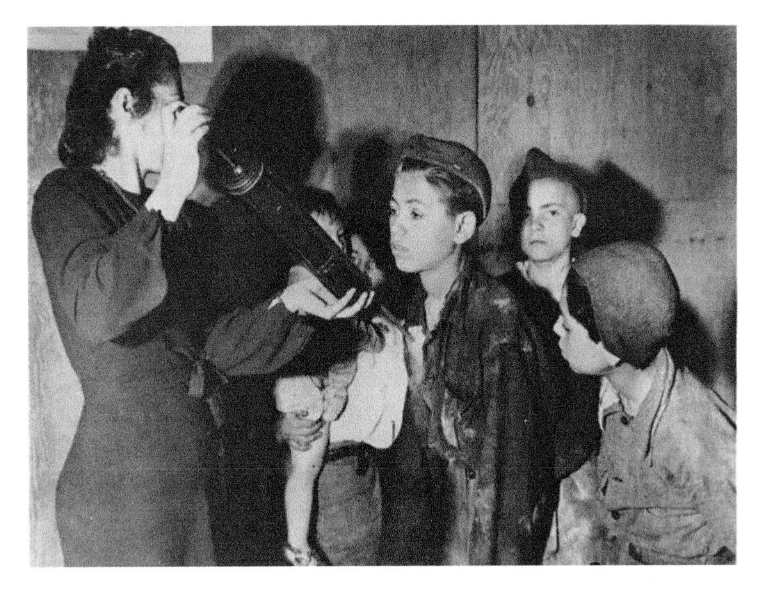

Fig. 1.3. DDT dust applied to children in Naples, 1944. (Photo courtesy of H.D. Hudson Manufacturing Co.).

programme and this resulted in an immediate resurgence of the vector. Only in countries such as the USA and Italy, with better housing, screens on doors and windows and a cold winter, was the transmission of malaria halted.

Fig. 1.4. Early design of aerosol 'bomb' containing DDT.

Very rapidly, DDT became used extensively to control agricultural pests, as it was very stable in sunlight. In the USA it was widely used to control cotton pests, notably the cotton bollworm. For added efficacy against the boll weevil, DDT was applied at 4 lb/acre but mixed with methyl parathion and toxaphene. Sprays were applied from aircraft, but with high temperatures in the cotton states, it is possible that some of the spray was not deposited on the crops; instead it was carried upwards on thermals and then distributed globally by the jet stream, returning to earth where snow fell, the persistent chemical being filtered out by the snow. As pointed out by Brown (1951), applications were not very efficient; he calculated that to control a population of 1 million mosquitoes per acre, only 30 mg of DDT would be needed for their total destruction (1 million $\times$ 3 $\times$ 10^{-2} µg), yet using the most modern and efficient method then available required 3000 times this amount (0.224 kg a.i./ha).

In Africa, cotton yields were very low, so Eric Pearson, the then Director of the Commonwealth Institute of Entomology established a Cotton Pest Research Project within the Federation of Rhodesia and Nyasaland in the 1950s. He wanted the entomological team to help small-scale farmers achieve yields similar to those obtained by farmers in Texas. Some trials in southern Rhodesia had already shown that yields could be increased with endrin sprays, but a programme of bioassays revealed that carbaryl was most effective against *Diparopsis castanea* and DDT was better against *Helicoverpa armigera*. This was confirmed by initial field trials; so starting in the 1960/61 season, a spray programme based on scouting to determine which bollworm was present was started on farms in both Rhodesia and Nyasaland. The DDT used was the 75% wettable powder developed to meet WHO specification, as the highly micronized powder (WP) applied as suspension was shown to be as effective as the emulsifiable concentrate (EC) formulation, presumably due to adhesion of the small particles on pubescent plant surfaces where the bollworms were walking. Clearly, very small particles adhered better to the plant due to van der Waals forces. Yields were improved from, generally, under 500 kg seed cotton/ha to over 1500 kg/ha where rainfall was adequate; and, later, yields over 3000 kg/ha were achieved under irrigation. Using a WP formulation of both carbaryl and DDT, these insecticides were packaged in sachets, each containing sufficient powder for one knapsack sprayer load, to prevent farmers having to measure out the small quantities required. With ground equipment, nozzles were positioned between the rows of cotton and the spray volume increased in relation to plant height from 50–200 l/ha; so when DDT was applied, the maximum dose was 1 kg a.i./ha in 200 l/ha. Thus, by more accurate distribution of spray, the dose applied was always less than some of the recommendations in the USA where DDT was relatively inexpensive. Dimethoate was added mainly as an acaricide, if red spider mite populations had to be controlled. Demeton-S-methyl (Metasystox) came on the market but was not recommended as it was too toxic compared to dimethoate. With higher yields, the area of cotton grown in Rhodesia increased, but finding 200 l of water to spray limited uptake by many small-scale growers in Nyasaland (now Malawi). Subsequently, an ultra low volume (ULV) spray technique was developed (see Chapter 2) but was used most extensively in francophone countries in west Africa.

The effects of using DDT were soon noted by environmentalists, and prompted Rachel Carson to write *Silent Spring* in 1962. There was particular concern about the persistence of DDT in the environment. Subsequently, DDT was banned by the Stockholm Convention along with other organochlorine insecticides as they were regarded as persistent organic pollutants (POPs), which were accumulated in the food chain and detrimental to the environment. By 2001, more than 120 countries had signed up to the Stockholm Convention. However, its use for indoor residual spraying was allowed to continue.

Lindane

The gamma isomer of benzene hexachloride, otherwise known as hexachlorocyclohexane (BHC), was named after van der Linden, who discovered the isomer in 1912. It was in 1942 that its insecticidal activity was noted (Busvine, 1964). The volatility of BHC in the field resulted in inadequate residual spray deposits, but allowed it to have a fumigant effect on insects in crevices when applied indoors, and controlled insects in tree canopies that were not easy to treat. It was used as the main insecticide to control cocoa pests (*Sahlbergella singularis* and *Distantiella theobroma*) in Ghana in 1954 (Stapley and Hammond, 1959), but within ten years the capsids resistant to γBHC were detected (Dunn, 1963). It was applied, initially, with a compression sprayer with a separate motorized fan to project the spray to the upper canopy, which led to the first motorized knapsack mistblowers. It has been used in a shampoo to treat scabies by killing the mites and their eggs, if other treatments were ineffective, but has to be washed out after no more than 12 hours.

Other organochlorines

Various other insecticides that were introduced in the 1940s and 1950s included some analogues of DDT – methoxychlor, DDD and DFDT – but these were not used on a large scale compared with chlordane, the earliest cyclodiene insecticide. Technical chlordane was developed by chance in the 1940s while looking for a by-product of synthetic rubber manufacturing. It contained five isomers, one of which was heptachlor (with seven chlorines to stabilize the cyclodiene ring), which was more insecticidal than chlordane. Chlordane was sold in the USA until 1988 and used to control termites in over 25 million homes. The half-life of chlordane can be up to 30 years, so it provided long-term protection. The most widely used cyclodienes were aldrin and dieldrin. Aldrin, named after the German chemist Kurt Alder, is not an effective insecticide, but it oxidizes to the epoxy dieldrin, which was a very effective insecticide. Both were used extensively in agriculture in the USA until about 1974, although dieldrin continued to be used for controlling termites until 1987.

Dieldrin was the insecticide selected for controlling locusts, especially the desert locust (*Schistocerca gregaria*) in Africa, using drift spraying to deposit spray on the vegetation being eaten by the locusts (Courshee, 1959). Shell, which marketed dieldrin, produced an excellent film called *The Rival World*, which showed the impact of locusts and efforts using aircraft to spray swarms. The Food and Agriculture Organization (FAO) decided to have stockpiles of dieldrin positioned in a number of African countries for emergency use, in case an upsurge of locusts occurred. Then, when a major outbreak occurred in west Africa in the 1980s, the USA demanded that the use of dieldrin should be discontinued due to its persistence in the environment. These then became obsolete stocks, which were extremely expensive to remove and incinerate. FAO developed a list of alternative insecticides for locust control so that countries had a choice. The organophosphate chlorpyrifos was often being applied, as it was readily available. FAO agreed that research was needed to find an effective biological control, which resulted in a major multi-country financed project, managed by CAB International, which developed the mycoinsecticide Green Muscle based on a *Metarhizium* species, later recognized as *M. acridum*. Although more expensive than an organophosphate insecticide, it had no effect on birds eating moribund locusts, and old stocks degraded, so there were no environmental costs or any need to process obsolete stocks.

Dieldrin was also used to control tsetse flies (*Glossina morsitans*) in northern Nigeria using a pressure-retaining compression sprayer to apply it to the resting sites of flies at the base of trees.

Another organochlorine insecticide was endrin, first produced in 1950, which was widely used on cotton and rice crops but subsequently discontinued in 1972. It was considered too toxic for small-scale farmers and was a POP. Toxaphene, also known as camphechlor, was introduced in 1947 by Hercules Inc. It was subsequently used in a mixture with methyl parathion and DDT on cotton as it was said to increase the persistence and effectiveness of methyl parathion and enhance the impact on boll weevils. It was used extensively in Nicaragua on cotton, applying as much as 31 kg/ha in 1985 (Carvalho *et al.*, 2003). It was persistent in soil, but in air the half-life was less than a day (Anon, 1977). It was banned in the USA in 1990 and globally by the 2001 Stockholm Convention on Persistent Organic Pollutants.

Endosulfan (Thiodan) is a chlorinated hydrocarbon, but differs from DDT as it is also an organic sulfite. It is more toxic to mammals than DDT and is now being phased out as it is considered to be an endocrine disrupter and too hazardous to use. Nevertheless, as a non-systemic insecticide/acaricide, it was used extensively in India. In Africa aerial applications were made against tsetse flies, while elsewhere sprays were used on a range of crops including cotton, potatoes, tomatoes and apples. In the UK there was a pre-harvest interval of six weeks when sprays were applied to blackcurrants to control blackcurrant gall mite (*Cecidophyopsis ribis* (Westw.)). When it was first provided for trials in Africa, spray operators immediately reported headaches. This problem was reported to the manufacturer who subsequently improved the product, presumably by removal of an isomer or contaminant introduced during the production process. Its use has been banned in many countries (see Chapter 9).

Organophosphates

Organophosphates (OP) are cholinesterase inhibitors that disable cholinesterase, an enzyme essential for the central nervous system to function. Chemists in Germany, such as Gerhard Schrader, had started to investigate organophosphates as insecticides in the 1930s and this led the government to get him to develop nerve gases such as sarin, tabun and soman as chemical weapons, although these were not used during World War II. However, after the war, chemical companies in the USA gaining access to Schrader's work and patents began synthesizing organophosphate insecticides. Parathion – *O,O*-diethyl-*O*-p-nitrophenyl thiophosphate, originally known by the code E-605, was the first product to be marketed as Folidol. The methyl analogue of parathion, methyl parathion has similar toxicity to mammals, but was claimed to be more active against some insects, including the boll weevil. Tetraethylpyrophosphate, called TEPP, was discovered as an aphicide in 1938, and *p*-nitrophenyl thionobenzene phosphonate, called EPN, was developed at the same time, but not widely used. A systemic OP was schradan, pyro-phosphoryl-tetrakis-diametylamide, which was marketed as Systox. Small sticks coated with schradan could be stuck into pots to protect young seedlings.

Later, a vast number of other organophosphate insecticides were developed, which exhibited a wide range of toxicities to mammals.

Malathion

Malathion, first reported in the USA in 1952, is far less toxic than parathion. According to the WHO classification of pesticides, the acute oral toxicity of parathion is 3–6 mg/kg (in class I), while malathion was unclassified with an acute toxicity to mammals of 1400 mg/kg. Malathion has been used extensively in public health against mosquitoes and other vectors of disease as well as on many crops, including sprays, with a protein hydrolysate or a yeast bait for fruit flies. There was one major problem among 7500 workers in Pakistan in 1976, when two poor-quality formulated products of malathion, containing isomalathion, were used to spray houses. Banning of these products, plus further training to reduce operator exposure, followed, and since then the problem has not recurred. Certain countries, especially in the Middle East, used technical malathion as a ULV spray for locust control. As the technical material is a liquid, it did not need to be formulated, so could be sprayed directly, but it was pointed out that a ULV formulation containing only a small amount of malathion could be just as effective.

Temephos

Temephos also has a low toxicity and has been used in rivers to control blackfly (*Simulium* spp.) larvae to reduce transmission of onchocerciasis. The Onchocerciasis Control Programme (OCP) operated in west Africa for two decades to break the transmission of the parasite causing river blindness.

Dimethoate

Dimethoate, which was introduced in 1951 by American Cyanamid as both an insecticide and acaricide, is readily absorbed and distributed through plant tissues and degrades quite quickly. It was regarded as much better to use than more toxic insecticide/acaricide demeton-S-methyl (Metasystox) introduced by Bayer in 1957. It was the concerns about the application of Metasystox on Brussels sprouts, even with a three-week pre-harvest interval (PHI), that led the UK government to set up a Working

Party on Precautionary Measures Against Toxic Chemicals in 1950, with Professor Zuckerman as chairman. This led to the Advisory Committee on Pesticides, renamed the Expert Committee in 2014. The Pesticide Safety Precautions Scheme (PSPS) operated from 1957 for agricultural products, and from the 1970s, non-agricultural products were added. As it was a voluntary scheme, there was pressure for a statutory system, which followed with the Food and Environment Protection Act (FEPA) in 1985. Later harmonization of pesticide legislation led to EU directives, discussed later (see Chapter 10).

Other early OPs included phorate, dichlorvos, trichlorfon, fenthion, menazon and phosphamidon, each having a specific role.

Phorate

Phorate (*O,O*-diethyl S-ethylthiomethyl phosphorodithioate) was marketed as a systemic insecticide and acaricide in 1954 as Thimet. It was used as a seed treatment as it gave up to eight weeks' control of sucking pests such as aphids, thrips and leaf hoppers.

Dichlorvos

Dichlorvos (2,2-dichlorovinyl dimethyl phosphate, commonly abbreviated as DDVP) was first marketed in 1959. Due to its vapour action it became widely used against household and public health pests, particularly when sold as a Vapona plastic strip to hang up in houses. Safety concerns have reduced its use and it has been banned in Europe since 1998. However, in the USA, a low dose of naled (Dibrom), dimethyl 1,2-dibromo-2,2-dichloroethylphosphate, is applied in an aerial spray to control mosquitoes; for example, in Florida during 2016 it was applied to combat *Aedes aegypti*, vector of the Zika virus. It is very volatile and breaks down to dichlorvos. This is rapidly dissipated, and specialists at both the CDC and EPA, as well as independent universities, argued that naled was safer than other chemicals and should not cause significant health issues due to the low level of exposure. A small piece of the Vapona plastic strip was very useful for entomologists operating light traps, as the insects were killed very quickly inside the trap.

Trichlorfon

Trichlorfon (dimethyl 1-hydroxy-2,2,2-trichloro ethanephosphonate) is a non-systemic insecticide, rapidly hydrolysed in plants. Dipterex was one of the trade names used. In Africa it was applied as a dust in the 'whorl' of maize leaves to control stem borers (Fig. 1.5)

Fig. 1.5. Applying Dipterex to young maize.

Fenthion

Fenthion (*O,O*-Dimethyl *O*-[3-methyl-4-(methylsulfanyl)phenyl] phosphorothioate) is perhaps known most for its use in controlling the weaver bird (*Quelea quelea*) in Africa, large colonies of which may comprise thousands of birds, causing considerable damage to cereal crops. Aerial sprays at dusk were effective when applied, as birds congregated at roosting sites. In some countries it is banned due to its impact on bird populations.

Menazon

Menazon (S-(4,6-diamino-s-triazin-2-yl) methyl *O,O*-dimethyl phosphorodithioate) was introduced as a selective aphid insecticide but was later replaced by pirimicarb.

Phosphamidon

Phosphamidon ((*E*/*Z*)-[3-Chloro-4-(diethylamino)-4-oxobut-2-en-2-yl]) dimethyl phosphate was marketed as Dimecron in 1956. It is a highly hazardous insecticide and is now included in the Rotterdam Convention (previously a voluntary procedure but in force since 2004) and requires prior informed consent (PIC) before it can be exported to a country.

Application of OP insecticides increased when the organochlorine insecticides were banned in the 1970s.

Carbamates

Carbaryl was introduced by Union Carbide under the trade name Sevin in 1958. It was in a screen searching for possible herbicides, when it was noticed that flies in a glasshouse had died. It inhibits cholinesterase in a similar way to organophosphates, but rapidly breaks down to α-naphthol and is excreted. Its acute mammalian toxicity is similar to DDT, so was evaluated in southern Rhodesia and shown to be highly effective against the red bollworm and cotton stainers (*Dysdercus* spp.). It was subsequently used extensively on cotton.

With a broad spectrum of activity, carbaryl (1- naphthyl *N*-methylcarbamate) became widely used throughout the world to control a range of pests, including fleas on pets. In the mid-1970s, in the USA, there was a request that the EPA consider carbaryl under the Rebuttable Presumption Against Registration (RPAR) procedure as it was considered that the chronic effect of exposure to carbaryl might be more hazardous than previously thought. However, it was later withdrawn from RPAR and some changes were made to the labelling.

In India, in 1984, at Union Carbide's factory in Bhopal, which manufactured Sevin, the chemical methyl isocyanate (MIC) was contaminated with water, releasing an extremely toxic gas that killed nearly 4000 people. Many more living in the slums close to the factory were severely affected by the gas and an estimated 15,000 died later, with many more continuing to suffer chronic symptoms. It was not until 2006 that the government confirmed that the leak had caused 558,125 injuries, including 38,478 temporary partial injuries and approximately 3900 permanently disabling injuries. Despite the enormity of the tragedy, seven ex-employees were sentenced in 2010 by an Indian court to only two years' imprisonment and a fine of $2000 for causing death by negligence.

Other Compounds

DNOC

DNOC (4,6-dinitro-*o*-cresol) was introduced by Bayer in 1892 and was assessed as an insecticide to control locusts in east Africa. In 1945, a 2.5% formulation in an oil spray was applied attacking flying swarms. On one occasion, a kill of some 3 million locusts was recorded following an application and this led to further work with a 20% formulation applied at 0.7–1.0 gallon/acre using the 'aerial curtain' method (Rainey and Sayer, 1953). Other insecticides replaced DNOC for locust control. The highly toxic DNOC was also listed as a selective herbicide that was used on grasses and as a defoliant on potatoes, but was banned in the UK in 1989 due to evidence of teratogenicity in related di-nitro compounds. Concerns about its safety are referred to in Chapter 9.

Metaldehyde

In 1936, in southern France, a solid fuel sold as meta-tablets was used by some campers. Some of the tablets were left on the ground and it was noticed that there were dead slugs in the same area. Thus the molluscicidal activity of metaldehyde was discovered. This cyclo-octane is approved in the EU; it kills the slugs by contact and stomach poisoning, which stimulates mucus/slime production resulting in desiccation of the slug.

Acaricides

A number of acaricides were commercialized in the 1950s. **Dicofol** (Kelthane) is chemically similar to DDT but is not as persistent in the environment. It has a broad range of activity against mites, but was not considered to be toxic to bees and beneficial predators. Another very effective acaricide, not toxic to bees, is **tetradifon** (4-chlorophenyl 2,4,5-trichlorophenyl sulfone) (Tedion), but it is no longer registered. **Amitraz** was an early acaricide used on cotton and other crops, but due to health concerns it is no longer registered.

Fumigants

DD was a mixture of dichloropropane and dichloropropene used as a fumigant to control plant parasitic nematodes, specifically the potato cyst nematode *Globodera rostochiensis* (eelworms), but was also used by tobacco farmers to control the root-knot nematodes in the seed beds before the seedlings were transplanted in the field. Ethylene dibromide was an alternative soil fumigant. These highly poisonous and volatile liquids were applied by injection into the soil to a depth of at least 20 cm to percolate through the soil and protect the young roots from the nematodes. Treated soil was covered by a plastic sheet to keep the gas within the soil for a period after application. Later, methyl bromide replaced these in many crops, although its use has been phased out under the Montreal Protocol as it is an ozone-depleting substance. Ethylene dichloride was a similar fumigant used to protect grain in storage (see also Chapter 6).

Herbicides

With the Industrial Revolution people moved from farming to factories, but two world wars also adversely affected the availability of workers to cultivate, weed and harvest farmers' crops. There was a need to control weeds chemically, so in seeking better weed control, ICI began, in 1936, to study the effects of plant hormones to determine whether weeds might be killed without harming crops. In 1942, the effect of increasing the dose of certain substituted phenoxy acids that normally stimulated plant growth was shown to affect plant growth to such an extent that the plants died. Thus when indole-3-acetic acid (IAA), the naturally occurring auxin, was used at high concentrations, it could stop plant growth (Templeman and Marmoy, 1940). Templeman and Marmoy published their finding in 1940 that IAA killed broadleaf plants within a cereal field. The first herbicide to be developed from this discovery was MCPA (4-chloro-2-methylphenoxy) acetic acid to control broad-leaved weeds in cereal crops. At the same time, in the USA, Pokorny (1941) was looking for a more stable phenoxy acid and this led to the synthesis of 2,4-dichlorophenoxyacetic acid (2,4-D) and 2,4,5-trichlorophenoxyacetic acid (2,4,5-T), both phenoxy herbicides and analogues of IAA.

MCPA

This became widely used either as the sodium or potassium salts, or both, in a mixture to control broad-leaved weeds in cereals, notably wheat, barley and oats. It was also used in grasslands to improve pastures. MCPB (4-chloro-2-methylphenoxy butryric acid) was more selective, where crops were undersown with clovers and certain leguminous plants that lacked the enzymes to convert it to MCPA.

2,4-D

Since the first commercial release of 2,4-D in 1945, it has continued to be a major selective and low-cost herbicide, controlling broad-leaved weeds in cereal crops. 2,4-D became widely used in the USA as a replacement for the hoe, applied as a sodium salt or as an amine or ester derivative after the cereal crop had fully tillered, but before shoots were present. A new choline salt version of 2,4-D (2,4-D choline) was developed much later as a less volatile herbicide (Peterson *et al.*, 2016) and is expected to be applied to crops that have been engineered to be tolerant to 2,4-D.

The use of effective broad-leaved weed herbicides resulted in a major improvement to wheat yields in the UK, which were enhanced by the breeding of semi-dwarf varieties less prone to lodging, which made harvesting easier. In Fig. 1.6, the crop suffered severe yield loss due to a prolonged drought in 1976, but later, greater use of fungicides enabled farmers to achieve much higher yields.

2,4,5-T

2,4,5-T was developed as it was more effective against woody weeds. It was present in Agent Orange with 2,4-D in approximately equal amounts of the n-butyl esters used by the US military to destroy forest cover during

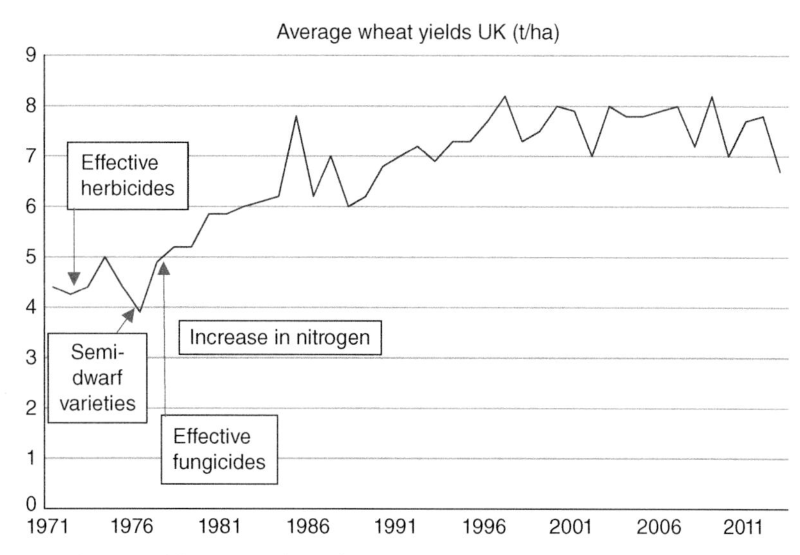

Fig. 1.6. Average UK wheat yields from 1971 to 2013. (Redrawn from Orson, NIAB.)

the Vietnam War (1961–1971), but its contaminant, 2,3,7,8-tetrachlorodibenzodioxin (TCDD), an extremely toxic dioxin compound, resulted in serious health effects. Since 1970, use of 2,4,5-T has been phased out, and in 1983, use in the USA ceased.

The volatility of these herbicides has been of particular concern in a number of situations. In the UK, a phenoxy herbicide applied to cereal crops during a warm May in 1976 led to considerable damage to vegetable crops in the Vale of Evesham. Initially, there was concern that it was due to spray drift, and this led to studies on drift (Elliott and Wilson, 1983) and development of a spray quality classification (Doble *et al.*, 1985), but later the movement of vapour from the sprayed crop was considered the cause of the damage (Thompson, 1983). Cotton crops grown near railway lines in Rhodesia were damaged due to sprays applied to the railway track as plants were very susceptible to 2,4-D. Farmers had to be careful to wash sprayers thoroughly after applying 2,4-D and similar herbicides, as trace amounts could easily damage other crops.

A number of other herbicides were developed in the 1950s and 1960s. These included dalapon, TCA, CIPC, barban, simazine, atrazine, amitrole, diuron, sodium chlorate and dacthal, but some are generally no longer manufactured or registered.

Dalapon

Dalapon (2,2-dichloroproprionic acid) was absorbed and translocated in grasses and has been used as the sodium salt to control perennial grasses, such as couch grass, Bermuda grass and Johnson grass. The major use of dalapon has been on sugarcane and sugar beets.

TCA

TCA (trichloroacetic acid) was introduced in 1947 and used as the sodium salt to control perennial grasses, being absorbed mainly through the roots.

CIPC

In 1951, CIPC, better known as chlorpropham (isopropyl (3-chlorophenyl) carbamate), was introduced as a plant growth inhibitor/herbicide, which was used as a sprout suppressant for grass weeds. One particular use has been to inhibit potato sprouting by allowing a fog to percolate through stacks of potatoes in stores.

Barban

Barban, a carbamate, was used as a post-emergent, selective herbicide for control of wild oats and other grasses, but is now obsolete.

Simazine

Simazine (2-chloro-4, 6-bis(ethylamino)-S-triazine) was developed in 1956 and used initially to control weeds on paths and railway tracks and on industrial sites to control germinating annual grasses and broad-leaved weeds.

Atrazine

Another triazine, atrazine (1-Chloro-3-ethylamino-5-isopropylamino-2, 4, 6-triazine), was developed at about the same time and was extensively used to control weeds in maize, which could detoxify it. In Nigeria, it was applied at very low volume sprays using hand-carried spinning disc sprayers known as 'herbis' (Fig. 1.7) or 'handys'. In the USA, atrazine was detected in drinking water and it was suspected of being an endocrine disrupter; so with considerable controversy about its effects in the environment, the EPA has been reviewing its use. In the EU, its use was banned in 2004, when groundwater levels exceeded the limits set by regulators.

Amitrole

Amitrole is a non-selective systemic triazole that was developed in 1953 to control a wide range of perennial grasses and broad-leaved weeds, especially non-crop lands. In the USA, one use for which it was registered was post-harvest use on cranberries. This led to very low residues on portions

Fig. 1.7. Spraying atrazine at International Institute of Tropical Agriculture.

of the cranberry crops of 1957 and 1959 when the authorities enforced the Delaney Clause, which prohibits any amount of a cancer-causing substance to be in or on food. This is thought to have prompted growers to read and follow pesticide label directions. It was used in a number of countries and is considered suitable for controlling glyphosate resistant weeds, although it is much slower acting than glyphosate. It is no longer registered in the UK.

Diuron

Diuron (3-(3, 4-dichlorophenyl)-1, 1-dimethylurea), introduced in 1954, is a very specific and sensitive inhibitor of photosynthesis, used as a total herbicide, particularly in non-crop areas and woody crops. It takes about a year to dissipate in soils.

Sodium chlorate

Use of sodium chlorate as a non-selective herbicide dates back to 1910. Residues can remain in the soil for up to five years, depending on rate of application and soil conditions. Although used mainly in non-crop land, it has been used as a defoliant and desiccant in some crops. Linuron, another urea herbicide similar to diuron, was developed in 1962 and is used to control annual meadow grass and broad-leaved weeds in potatoes and other vegetable crops. In 1955, naptalam-sodium was introduced as a pre-emergence herbicide used for the control of a wide range of weeds and grasses in vegetable, fruit and non-food crops. It is no longer manufactured.

Dacthal

Dacthal (DCPA) was first registered for use in the USA in 1958. It is a pre-emergent herbicide that kills grass and many common weeds without killing sensitive plants such as flowers, fruits, vegetables, turf and cotton. Production continued until 1998, but the product was reintroduced in America in 2001.

More on herbicides is to be found in Chapter 4.

Fungicides

Bordeaux mixture has already been mentioned and was prepared by mixing 5 lb of copper sulfate in 25 gallons of water and, separately, 5 lb of fresh hydrated lime in 25 gallons of water; then the two were mixed together. Other copper-based formulations soon followed. Burgundy mixture (1887) used sodium carbonate in place of calcium hydroxide, 'Eau Celeste' (1885) utilized ammonium hydroxide, and in Cheshunt mixture (1890) the base was ammonium carbonate. In each case the copper is complexed on the leaf surface, which prevents overt phytotoxicity but is slowly leached at a sufficient concentration to inhibit spore germination. None of these, nor other copper compounds (see below), significantly replaced Bordeaux mixture, which was a huge advance at the time and is still an effective fungicide by modern standards. The freshly prepared mixture, as described above, forms a gelatinous precipitate, which adheres well to plant surfaces and also undergoes local redistribution and improves coverage. Other copper-based compounds include copper oxychloride, copper oxinate and cuprous oxide. Boots Pure Drug Co. introduced a finely divided copper oxychloride suspension, patented in 1931, which became popular as a substitute for Bordeaux mixture. It had the advantages of not requiring mixing in the field and not being corrosive to steel. These fungicides were used to control blister blight on tea, leaf rust on coffee, late blight on potatoes and leaf spot on bananas, as well as the original target, vine downy mildew.

Bordeaux mixture was used for many years in west Africa against black pod disease of cocoa where, unfortunately, its low cost allowed its use to excess, with adverse environmental side effects as well as, reputedly, sometimes contributing a blue tinge to plantations. Toxicity to non-target organisms such as soil invertebrates has curtailed the use of copper compounds, although small quantities are probably beneficial, copper being a plant micro-nutrient. Perversely, despite containing a heavy metal, Bordeaux mixture was allowed for some time on organic farms on the basis of it being a 'traditional' product. It is interesting to speculate that if the first effective foliar fungicide (Bordeaux mixture) had not been such a good product, the history of disease control might have taken a very different route. There is a direct parallel with medicine: if the first antibiotic, penicillin, had not been so remarkably effective, chemotherapy in medicine might never have developed as it has.

Other inorganic fungicides have included sulfur, discussed above, and mercury salts. Both mercurous and mercuric chlorides received some use for treating bulbs and corms pre-planting and against common scab of potato, club root disease of brassicas and on turf grass. However, it is organic forms of mercury that have been used much more extensively and with considerable success in disease control. Compounds with the general formula R-Hg-X, where R is an aryl or aromatic group and X is an anion – for example phenyl mercuric acetate – proved excellent at controlling bunt (covered smut) of cereals and also gave very good control of other seed-borne diseases of cereals such as oat leaf stripe and seedling blights caused by *Pyrenophora* and *Cochliobolus* species. These chemicals were too phytotoxic for use on foliage but had a long period of use as seed dressings under such trade names as Uspulam (Bayer) and Ceresan (ICI). Formulations included dyes to obviate against food use and repellants to deter seed-eating birds, but the hazards of cumulative environmental contamination led to their phasing out during the 1980s. Tin is the only other heavy metal to have seen use as a fungicide, exclusively as organo-tin compounds, similar to those of mercury. Phenyl tin (Fentin) acetate and hydroxide have seen use against leaf spots and potato late blight, usually as the final spray of the season because of mild phytotoxicity and also efficacy against tuber infection caused by zoospores washed down into the soil.

Other protectant fungicides, even where introduced prior to 1960, are considered in Chapter 5.

New chemistry led to the first organic fungicide, thiram (tetramethylthiuram disulfide), being developed. It was used primarily as a seed treatment, aimed at controlling damping off of seedlings and other seed-borne pathogens. It was also applied as a foliar spray to control apple scab (Fig. 1.8) and botrytis on strawberries. Another new fungicide was captan – N-trichloromethylmercapto-4-cyclohexane-1,2-dicaroximide – a foliar fungicide which was applied on apples at 10–14-day intervals after the buds had burst; the shorter interval being recommended in wet weather or if the trees made rapid growth, also influenced by the variety being sprayed. Another new protective fungicide developed at the same time was zineb – zinc ethylenebisdithiocarbamate – marketed as Dithane Z78. It was used on several vegetable crops to control blight, mildew and rusts. Dithane M-22 was a mixture of maneb (manganese ethylenebisdithiocarbonate) and manzate and used as a spray or seed dressing. Folpet, a phthalimide, was introduced in 1952 and remains registered as a multi-site protectant in the EU. It controls mildew on fruits, vegetables and ornamentals. Dodine, a guanidine first reported in 1957, controls scab, leaf spot and other foliar diseases. It is also used as an effective algal growth inhibitor in non-agricultural applications. Its registration in the UK is due to expire in 2021.

Fig. 1.8. Unidentified man spraying an apple tree. (Photo courtesy of Wenatchee Valley Museum & Cultural Center, 83-15-180.)

A number of organic mercury fungicides were marketed as seed dressings in the UK from 1932 to suppress bunt, but also had an impact on oats by controlling leaf stripe, thought previously to be due to bad weather (Ordish, 1952). They were extremely poisonous, so their use was discontinued.

Fentin hydroxide was used in the late 1950s but is no longer registered.

Thus, prior to 1960, the number of pesticides had expanded to combat the increasingly realized extent of crop losses. Ordish (1952) provided estimates of crop losses in several European countries, the USA and parts of Latin America, indicating extremely high economic losses of important crops prior to 1950. In the following chapters further consideration is given to the major new products and the problems associated with their use (see also Chapter 5).

Biopesticides

In the period immediately after World War II, the first biopesticide to be produced on a commercial scale was thuricide, containing the bacterium

Bacillus thuringiensis. The activity of the bacterium had been noted in 1911 when it was isolated from diseased larvae of the Mediterranean flour moth. In 1951, Professor Steinhaus, at the University of California, continued research on it. It was stated that it had to be applied so that the minimum ingestion rate was 450 viable spores/mg of insect larval body weight. It was tried on a number of crops but was less effective than many of the new chemicals, probably because of the need for the spores to be ingested. A few keen enthusiasts supported the use of a biopesticide, but in the 1960s they were generally less effective than other pesticides. Translating a biologically active substance in the laboratory to a commercial product would require considerable research and development, as witnessed by the ten-year programme needed to translate a metarhizium, later identified as *M. acridum*, from just being a fungus to being a viable insecticidal product, very effective against locusts and other acridids. Until recently, the market for biopesticides has been no more than 1% of the agrochemical market; thus it is rather understandable that the companies invested very little into the development of biopesticides.

References

Anon (1977) *The biologic and economic assessment of Toxaphene. A report of the Toxaphene assessment team to the rebuttable presumption against registration of Toxaphene*. Submitted to the Environmental Protection Agency on 12 September 1977. United States Department of Agriculture, Washington, DC.

Berry, C.L. (1990) 17th Bawden Lecture – The Hazards of Healthy Living: The Agricultural Component. *Proceedings of the Brighton Crop Protection Conference (Pests and Diseases)* 1, 3–13. BCPC, Farnham, UK.

Brown, A.W.A. (1951) *Insect Control by Chemicals*. Wiley, New York.

Busvine, J. (1964) The insecticidal potency of r-BHC and the chlorinated cyclodiene compounds and the significance of resistance to them. *Bulletin of Entomological Research* 55, 271–288.

Carson, R. (1962) *Silent Spring*. Houghton Mifflin Co., Cambridge, Massachusetts.

Carvalho, F.P., Montenegro-Guillén, S., Villeneuve, J.-P., Cattini, C., Tolosa, I., Bartocci, J.M., Lacayo-Romero, M. and Cruz-Granja, A. (2003) Toxaphene residues from cotton fields in soils and in the coastal environment of Nicaragua. *Chemosphere* 53, 627–636.

Courshee, R.J. (1959) Drift spraying for vegetation baiting. *Bulletin of Entomological Research* 50, 355–370.

Dalu, T., Wasserman, R.J., Jordaan, M., Froneman, W.P. and Weyl, O.L.F. (2015) An assessment of the effect of rotenone on selected non-target aquatic fauna. *PLOS ONE 10(11)*. Available at: http://dx.doi.org/10.1371/journal.pone.0142140 (accessed 23 March 2018).

De Ong, E.R. (1924) Toxicity of nicotine as an insecticide and parasticide. *Industrial & Engineering Chemistry* 16, 1275–1277.

Doble, S.J., Matthews, G.A., Rutherford, I. and Southcombe, E.S.E. (1985) A system of classifying hydraulic nozzles and other atomisers into categories of spray quality. *Proceedings of the Brighton Crop Protection Conference – Weeds*, 1125–1133.

Dunn, J.A. (1963) Insecticide resistance in the cocoa capsid, *Distantiella theobroma* (Dist.). *Nature* 199(4899), 1207.

Elliott, J.G. and Wilson, B.J. (1983) The influence of weather on the efficiency and safety of pesticide application: the drift of herbicides. Brighton Crop Protection Conference, Occasional publication 3.

Elliott, M., Farnham, A.W., Janes, N.F., Needham, P.H., Pulman, D.A. and Stevenson, J.H. (1973) A photostable pyrethroid. *Nature* 246, 169–170.

Glynne-Jones, A. (2001) Pyrethrum. *Pesticide Outlook*, 195–198.

Gnadinger, C.B. (1936) *Pyrethrum Flowers* (2nd edn). McLaughlin Gormley King, Minneapolis, Minnesota, USA.

Hayes, W.J. (1959) Pharmacology and toxicology of DDT. In: Muller, P. and Simmonds, S.W. (eds) *DDT, the Insecticide Dichlorodiphenyltrichloroethane and Its Significance. Human and Veterinary Medicine* vol. II. Birkhauser Verlag, Basel und Stuttgart, Germany.

Jefferies, P.R., Toia, R.F., Brannigan, B., Pessah, I. and Casida, J. (1992) Ryania insecticide: analysis and biological activity of 10 natural ryanoids. *Journal of Agricultural and Food Chemistry* 40, 142–146.

Kennedy, J. (1947) The excitant and repellent effects of mosquitoes of sub-lethal contacts with DDT. *Bulletin of Entomological Research* 37, 593–607.

Killeen, G.F., Fillinger, U., Kiche, I., Gouagna, L.C. and Knols, B.J.G. (2002) Eradication of *Anopheles gambiae* from Brazil: lessons for malaria control in Africa. *The Lancet Infectious Diseases* 2, 618–627.

Kitron, U. and Spielman, A. (1989) Suppression of transmission of malaria through source reduction: anti-anopheline measures applied in Israel, the United States, and Italy. *Reviews of Infectious Disease* 11, 391–406.

Lodeman, E.G. (1896) *The Spraying of Plants*. Macmillan, London.

Missiroli, A. (1948) *Anopheles control in the Mediterranean area*. Fourth International Congress on Tropical Medicine, Washington.

Ordish, G. (1952) *Untaken Harvest*. Constable, London.

Pampana, E.J. (1951) Lutte antipaludique par les insecticides a action remanente – Resultats des grandes campagnes. *Bulletin of the World Health Organisation* 3, 557–619.

Peterson, M.A., McMaster, S.A., Riechers, D.E., Skelton, J. and Stahlman, P.W. (2016) 2,4-D past, present, and future: a review. *Weed Technology* 30, 303–345.

Pokorny, R. (1941) New compounds: some chlorophenoxyacetic acids. *Journal of the American Chemical Society* 63, 1768.

Rainey, R.C. and Sayer, H.J. (1953) Some recent developments in the use of aircraft against flying locust swarms. *Nature* 172, 224–228.

Roark, R.C. (1933) Rotenone. *Industrial & Engineering Chemistry* 25, 639–642.

Roberts, D., Tren, R., Bate, R. and Zambone, J. (2010) *The Excellent Powder – DDT's Political and Scientific History*. Dog Ear Publishing, Indianapolis, Indiana, USA.

Simmonds, S.W. (1959) The use of DDT insecticides in human medicine. In: Muller, P. and Simmonds, S.W. (eds) *DDT, the Insecticide Dichlorodiphenyltrichloroethane and Its Significance. Human and Veterinary Medicine*, vol. II. Birkhauser Verlag, Basel und Stuttgart, Germany.

Smith, D.N. and Cooper, J.F. (1997) Control of powdery mildew on cashew in Tanzania using sulphur dust – an audit of sulphur fate and a proposal for a new dusting strategy. *Crop Protection* 16, 549–552.

Soper, F.L., Davis, W.A., Markham, F.J. and Riehl, L.A. (1947) Typhus fever in Italy 1943–45, and its control with louse powder. *American Journal of Hygiene* 45, 305–334.

Stapley, J.H. and Hammond, P.S. (1959) Large-scale trials with insecticides against capsids on cacao in Ghana. *Empire Journal of Experimental Agriculture* 27, 343–353.

Staudinger, H. and Ruzucka, L. (1924) Insektentotende Stoffe I–VI, VIII–X. *Helvetica Chimica Acta* 7, 177–259, 377–390 and 406–458.

Sullivan, A. (1938) Relative knockdown value of Pyrethrins I and II. *Soap Sanit Chemicals* 14, 101–105.

Tattersfield, F. (1931) Pyrethrum flowers – a quantitative study of their development. *Annals of Applied Biology* 18, 602–635.

Templeman, W.G. and Marmoy, C.J. (1940) The effect upon the growth of plants of watering with solutions of plant-growth substances and of seed dressings containing these materials. *Annals of Applied Biology* 27, 453–471.

Thompson, N. (1983) Diffusion and uptake of chemical vapour volatilising from a sprayed target area. *Pest Management Science* 14, 33–39.

Ujváry, I. (1999) Nicotine and other insecticidal alkaloids. In: Yamamoto, I. and Casida, J. (eds) *Nicotinoid Insecticides and the Nicotinic Acetylcholine Receptor.* Springer, Tokyo, pp. 29–69.

Wigglesworth, V.B. (1955) The mode of action of DDT. In: Muller, P.and Simmonds, S.W. (eds) *DDT, the Insecticide Dichlorodiphenyltrichloroethane and Its Significance.* Birkhauser Verlag, Basel und Stuttgart, Germany.

Young, S.J., Gunning, R.V. and Moores, G. (2005) The effect of piperonyl butoxide on pyrthroid-resistant associated esterases in *Helicoverpa armigera* (Hubner) (Lepidoptera: Noctuidae). *Pest Management Science* 61, 397–401.

2 Application of Pesticides

'In crop protection, the chemical weapon must be used as a stiletto, not as a scythe' (Brown, 1951) – a very sensible statement; yet attention to pesticide application has been limited compared with the huge investments seeking new chemicals. In the earliest attempts to apply a pesticide, the spray liquid was simply allowed to flow by gravity from a knapsack tank through a hose to a 'sprinkler' held above the crop. As pointed out earlier, once Bordeaux mixture had been developed, there was a lot of activity building knapsack sprayers with a manually operated pump. Soon larger equipment, consisting essentially of a large barrel fitted with a manually operated piston pump, was made and mounted on horse-drawn carts, some with an elevated platform for tree crops. A number of hydraulic nozzles were designed by manufacturers in France and the USA.

In 1858, the 'fantail' was introduced (Fig. 2.1a), projecting a solid stream on a flat surface, later developed as a 'diffuser' nozzle, such as the Vigoureux nozzle, which ultimately evolved into a 'deflector' nozzle. Another early nozzle, made in 1875, was known as a 'graduating spray' nozzle, which could vary from a solid stream to a fine spray. The 'gem' nozzle (Fig. 2.1b) was therefore the first variable cone nozzle and became widely used on knapsack sprayers over 100 years later. Another version had a wire gauze instead of a flat plate. This was followed by the 'boss' nozzle, which incorporated a stopcock that enabled the user to adjust the opening and the flow rate. A nozzle with an elbow and a needle to push a blockage through the orifice was developed, as early nozzles were easily clogged. Vermorel in France was the first company to manufacture this nozzle, which was referred to as the Vermorel nozzle (Fig. 2.1c) (Lodeman, 1896; Illustrations of nozzles used prior to 1900 in Lodeman's book have been re-drawn (Figs 2.1 a–c)).

Very little changed over the next 50 years (Fig. 2.1d); application equipment was manufactured by relatively small companies; thus, in the UK in the 1950s, there were many such companies, notably Allman and Co.,

Fig. 2.1a. 'Fantail' nozzle introduced *c.*1850.

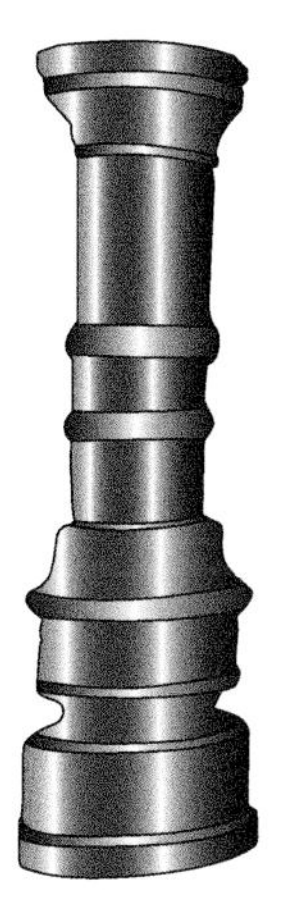

Fig. 2.1b. 'Gem' cone nozzle.

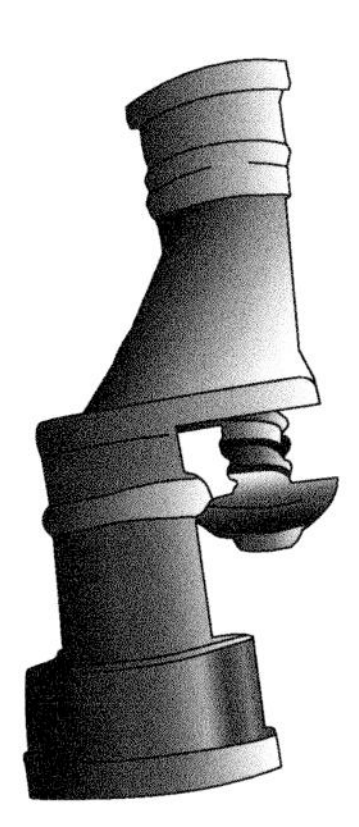

Fig. 2.1c. Vermorel nozzle.

Fig. 2.1d. Sprayers on sale in 1972.

Cooper Pegler, Dorman (now Team Sprayers), Drake and Fletcher, Evers and Wall, Four Oaks, W.T. French, Horstine Farmery, Kent Engineering and Foundry (KEF), Kestrel, and, a newcomer in 1954, Micron Sprayers. Older companies, such as Four Oaks, established in 1902, also supplied the coalmines with sprayers to apply whitewash. W.T. French became Associated Sprayers Ltd, but suffered a factory fire in 1965 and ultimately became Hozelock. Others merged with European companies, such as Hardi, and discontinued sprayer production or closed, so that by 2000, of the companies listed above, only Micron remained as a sprayer manufacturer in the UK. Horstine Farmery was the supplier of equipment to apply granules, a speciality now with Techneat. However, a few new companies began construction of larger self-propelled sprayers or sprayers to fit tractors. These included Bateman, Chafer (who acquired Horstine), Househam, Knight Farm Machinery Co. and Martin Lishman, with others importing spray equipment from Berthoud, Hardi, Technoma and other companies.

Dust Application

Application of dusts remained popular, even from aircraft, despite the inevitable drift of the finer particles. Simple hand dusters – the Cadet and, later, the Admiral Hudson – were used in Naples to treat people with a 10% DDT dust to arrest a typhus outbreak (Soper *et al.*, 1947). Bellows-type dusters were originally designed to apply sulfur dusts in vineyards. Another design was to have a hand-operated rotary blower mounted below a container for the dust, which could be mounted on the back of the operator. Often they were chest mounted and carried by a wide strap-like hook over the shoulders, despite the obvious hazard of the fine particles of dust being able to be inhaled by the operator. Apart from small spot treatments, dust application, even with large equipment, soon declined due to difficulties in the distribution and retention of dust particles on plant surfaces and risk of downwind drift (Figs. 2.2 a–d). The main developments from the 1950s were with spray applications, although some pesticides were applied into the soil as granules. This was particularly the case with WHO class I insecticides such as aldicarb, used for nematode control and early season control of sucking pests. Key to spray application was to achieve better delivery of the pesticide in terms of reduced volumes, better droplet spectra and improvements in formulation of the pesticide to fit the equipment suitable for different crops and other targets.

Nozzle Design

In 1946, Ransomes, Sims and Jeffries manufactured a boom sprayer with pneumatic nozzles, developed by ICI Plant Protection as the 'Agro' Atomiser sprayer. It could be fitted to a Fordson tractor. Each nozzle had a venturi so that liquid from a simple tube with a large opening was atomized

Fig. 2.2a. Applying a dust on cotton in Rhodesia in 1959, early morning, using a motorized mistblower. The dust did not deposit well.

Fig. 2.2b. Adaptation of mistblower with holed long tube. It was used to apply dust on rice in Japan.

by air from the venturi. Much later, a number of twin-fluid nozzles, such as the Airtec nozzle, were marketed and allowed farmers to reduce spray volumes applied (Jensen *et al.*, 2001). In the 1950s, some ceramic nozzles from Bray, designed for metering gas in domestic cookers, were used on tractor booms.

It was due to considerable research on atomization in other industries that led to changes in nozzle design for agricultural applications. Fraser (1956) recognized that in agriculture it had been usual to distribute the active ingredient by spraying dilute solutions through nozzles at high flow rates (i.e. up to 1200 gph per nozzle (91 l/min)) in conditions that may vary widely. He pointed out that spraying to 'run-off' usually wastes 95% of the

Fig. 2.2c. Applying sulfur dust on cotton using a stick to hit a bag of dust.

Fig. 2.2d. Bellows duster.

liquid and often does not completely wet the plants. Fraser concluded that there was a need to atomize and distribute very small volumes of up to 0.15 l/min in small drops and with greater uniformity over large areas of land.

Although sprayer manufacturers did produce their own nozzles, it was clear that nozzle manufacture was concentrated within a few specialist companies. Spraying Systems and Delavan in the USA, Lurmark (now Hypro) in the UK, Lechler in Germany and Albuz in France remain the key nozzle manufacturers for agriculture. For boom sprayers the main type has been the fan nozzle that was initially made from brass but is now almost entirely made with polyacetal, a semi-crystalline thermoplastic with high mechanical strength and rigidity. Stainless steel and ceramic nozzles of the same design have also been marketed to resist erosion by small sand particles in the water.

A range of hydraulic nozzles could be obtained with cone and fan sprays emitting different flow rates and spray angles. Initially, spray nozzles gave a spray angle of 80°, but the narrower 65° ones were also available; and later, 110° fan nozzles became the most popular for large areas of cereal crops, such as wheat being treated with wider booms, and farmers endeavoured to reduce spray drift by keeping the boom no more than 0.5 m above the crop. Narrower angles were needed where spray was projected to more specific targets, such as efforts to spray cocoa pods, while a very wide angle enabled a wide swath to be treated with a single nozzle when spraying herbicides around trees in palm oil plantations.

Flat fan (FF) nozzles on horizontal booms for arable crops were available in 005–08 sizes. The original numbering system was based on the output of a nozzle at 40 psi pressure in gallons/min, thus a 03 gave 0.3 gallons/min (1.2 l/min at 3 bar pressure). With the introduction of the polyacetal nozzles it was possible to colour code the different sizes of the fan nozzles, which follow an international standard (Table 2.1). Flow rate increases not only by having a larger orifice in the nozzle but also by increasing the pressure. Where spray volumes were high, the trend was to reduce the volume of water required to 200 l/ha. This has been widely

Table 2.1. Colour coding of hydraulic fan nozzles.

Nozzle size	Colour	Output in l/min at 3 bar pressure
01	Orange	0.40
015	Green	0.60
02	Yellow	0.80
025	Lilac	1.00
03	Blue	1.20
035	Brown red	1.40
04	Red	1.60
05	Brown	2.00
06	Grey	2.40
08	White	3.20
10	Light blue	4.00

followed, although even lower volumes, increased tractor speed and angling the spray are practices being adopted, with the choice of nozzle geared towards optimizing coverage. Some fan nozzles have two orifices spraying at different angles towards the crop to minimize spraying the soil between plants and to increase the amount deposited on the stem and foliage of the crop. Some manufacturers have angled caps to which standard hydraulic nozzles can be fitted.

The alternative to the fan nozzle with an elliptical orifice is the deflector nozzle, also referred to as an impact nozzle, with a round orifice directing the spray onto a flat surface, i.e. a development of the 'fantail' nozzle design of 100 years earlier. The shape of the orifice means that it is less liable to blockage with the lower flow rates and was promoted initially with the Polijet nozzle used on knapsack sprayers to apply herbicides. Later, the TurboTeeJet (TT) range redesigned the deflector nozzle so that the sheet of spray was directed downwards from the boom and not, as on a hand-carried lance, at about 70° to the lance.

All hydraulic nozzles produce a range of droplet sizes due to the way the liquid sheet emerges from the small orifice and breaks up into droplets. The smaller the orifice and the higher the pressure, the percentage of very small droplets increases. There is therefore a problem that the very small droplets are most likely to remain airborne and could be carried by the wind or by upward thermal airflow to areas outside of the crop. Spray drift was a big issue in the UK in the 1970s, although the main culprit was a volatile herbicide that resulted in vapour drift damage on vegetable crops miles away from where cereals were treated.

Fortunately, at this time, developments in measuring the droplet size of fuels and spray clouds had led to instruments using laser light to measure the spectrum produced by agricultural nozzles. A study of droplet spectra measured in laboratories in Europe and the USA led to the British Crop Protection Council (BCPC) classification of spray quality, using reference nozzles to counter differences between measurements obtained with the laser instruments. Spray quality was originally arranged with very fine, fine, medium, coarse and very coarse categories, but was later extended by the Spray Drift Task Force, set up in the USA, which added 'extra coarse'. For those involved in vector control using very fine sprays, a distinction was made between fogs and mists. The BCPC also introduced a common system of identifying nozzles irrespective of the manufacturer's nomenclature. Thus a nozzle with a spray angle emitted l/min at a set pressure. For example, FF110/1.2/3.0 indicated a 110° flat fan nozzle spraying 1.2 l/min at 3 bar pressure. On cereal crops, the widespread use of 110° nozzles is being challenged where farmers want to travel faster and have the boom slightly higher above the crop. As the 110° nozzle produces a finer spray, consideration is now given to narrower-angle nozzles, although more are needed across the width of the boom. The dilemma for farmers is how to reduce the quantity of water needed, avoid too fine a spray and still have sufficient coverage of the crop/weeds to obtain economical control of the pest. Further details on nozzles and method of application are given in Matthews *et al.* (2014).

Box 2.1. Droplet size, spray volume and spray retention

When sprays were applied at high volume, there was little attention given to spray quality, but once downwind drift of volatile herbicides was seen to damage sensitive crops, there was interest in defining the quality of sprays. Choosing certain nozzles, such as air induction nozzles, can reduce the amount of spray liable to drift, but in the quest for reducing drift, the need to optimize droplet retention on the crop tends to have been forgotten. In the 1950s, Ron Amsden had pointed out that the foliage of many crop plants is extremely hydrophobic. Peas and cabbages were the principal examples that were very difficult to wet, but applying smaller droplets or adding a surfactant could increase wetting and thus retention. Recent studies using high-speed photography confirm earlier data (Brunskill, 1956) that droplets of approximately 100 μm are more easily retained on a plant surface, whereas droplets >200 μm may bounce off or shatter on impact, with most bouncing off the leaf surface, leaving a small fraction of the spray on the surface. The loss of most of the spray in large droplets results in contamination of the soil, transferred by rain to ditches and rivers, hence the current need for untreated buffer zones to minimize the quantity that contaminates streams and, ultimately, rivers. Theoretically, by putting an electrostatic charge on droplets, retention on foliage can be increased, but this is inevitably on the nearest earthed surface. Thus good spray coverage is obtained with isolated single plants; but with row crops it is primarily the exposed upper foliage that is treated, while pests in the lower part of the canopy are not sprayed. Air assistance may be beneficial where there is space between plants or branches to allow penetration of the spray.

In the 19th century, Lodeman said that a fog was ideal, but it was necessary to control where the small droplets went rather than allow the wind to disperse them beyond the

Fig. 2.3a. Cascade impactor to collect small droplets.

Continued

Box 2.1. Continued.

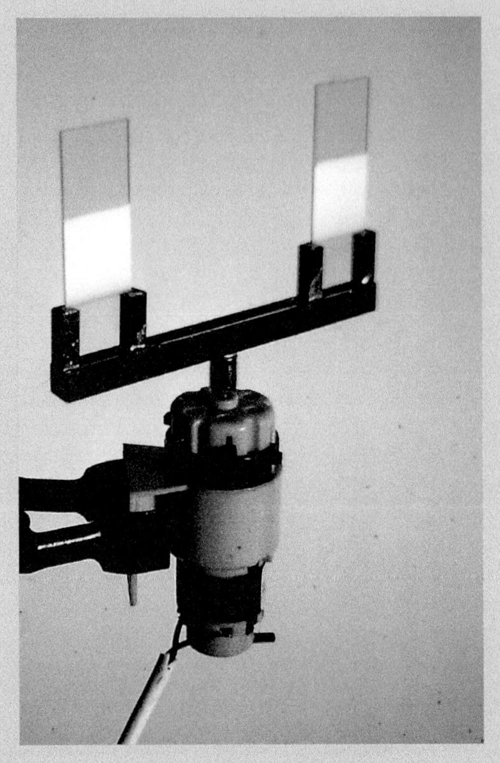

Fig. 2.3b. Rotating microscope slides treated with magnesium oxide to collect spray droplets.

crop boundary. Optimizing low- or very low-volume sprays with small droplets remains a challenge in the 21st century.

Measuring spray droplets

In the early days of modern pesticides, the measurement of small droplets was a laborious task. Droplets were captured on glass microscope slides, usually coated with a layer of magnesium oxide (MgO). A droplet hitting the surface would slightly flatten and leave a crater with a diameter 1.15 times wider than the droplet – the difference being referred to as the spread factor (May, 1950). The magnesium oxide layer needed to be sufficiently thick to avoid the larger droplets hitting the glass. Very small droplets were difficult to detect unless the spray liquid contained a dye to show up on the white surface. To sample the smallest droplets the slides are rotated at speed and also a narrower surface was used, for example by using rotorod samplers.

The development of computers and lasers resulted in several systems that can detect droplets passing through a beam of laser light and, depending on the equipment and its software, measure the droplet size and velocity of the droplets.

Continued

Box 2.1. Continued.

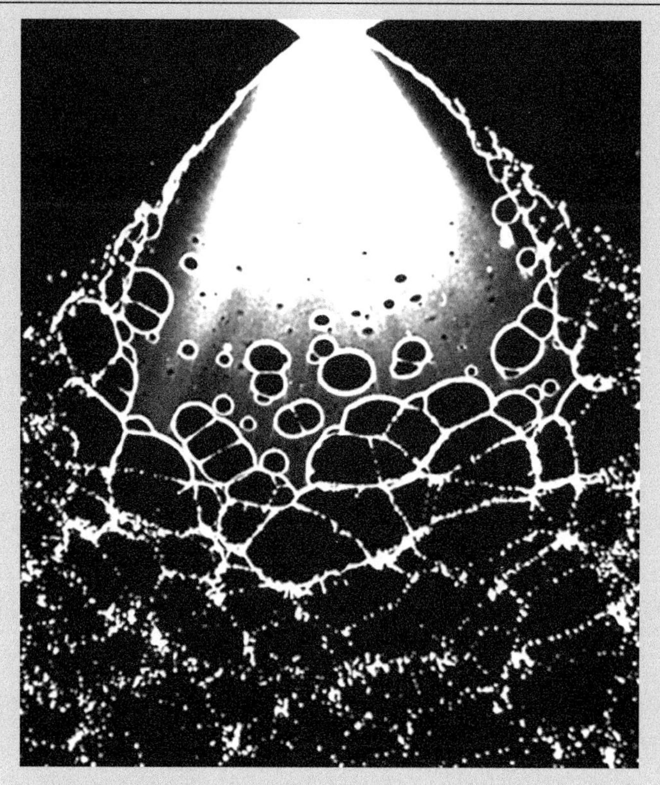

Fig. 2.3c. Spray sheet from a fan nozzle showing break-up of the spray into ligaments and droplets. (Photo from Fraser, 1956.)

Fig. 2.3d. Measuring spray droplets using a laser system.

Continued

Box 2.1. Continued.

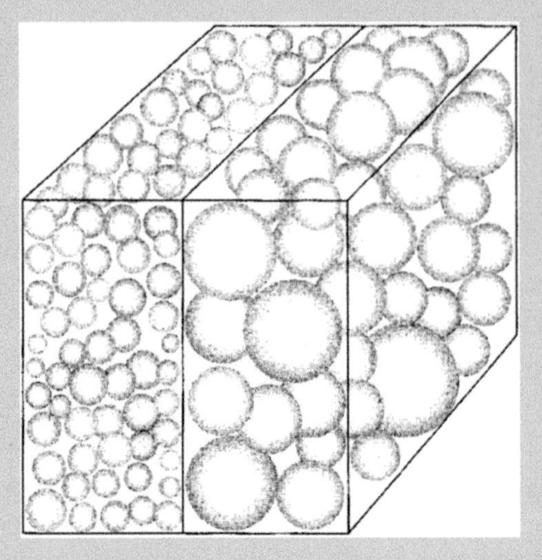

Fig. 2.3e. Diagrammatic representation of the volume median diameter (VMD) with half the volume having droplets smaller and half having droplets larger than the VMD.

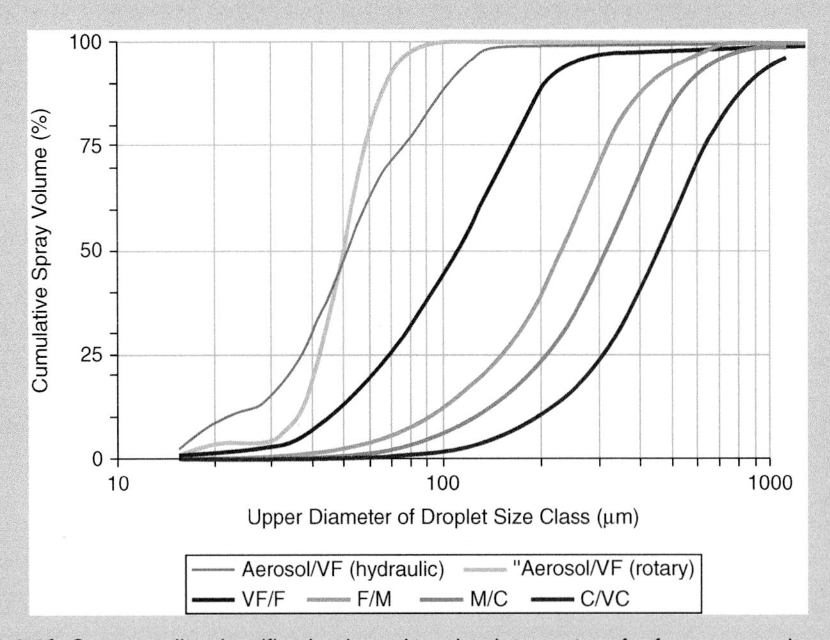

Fig. 2.3f. Spray quality classification based on droplet spectra of reference nozzles.

In the 1970s, in the USA, a foam nozzle was designed to apply spray with a much higher concentration of surfactant, but it did not receive general acceptance as the foam could be blown by the wind. However, later, much the same nozzle was designed merely to add air to the spray liquid. With the usual levels of surfactant in a spray, measurement of the droplet spectrum of this new nozzle revealed that the percentage of very fine droplets was significantly lower than with standard fan nozzles. It was subsequently marketed as the BFS BubbleJet, as many of the droplets contained an air bubble. This type of nozzle was manufactured by others and became known as an air-induction (AI) nozzle. It became popular as it reduced potential drift down-wind of a treated field. An assessment of potential drift reduction is possible using a large wind tunnel in which it is possible to collect dye solution sprayed from the test nozzle onto a series of horizontal plastic 'strings' placed at various distances (2–7 m) down-wind of the nozzle at a set height(s). The concentration of dye can be analysed to calculate the amount of spray collected at different distances down-wind and compared with data using a flat fan reference nozzle (FF/1.2/3.0) (Miller, 2014).

In contrast to the standard fan nozzles, the manufacture of AI nozzles varies as each company has its own configuration of allowing air to mix with the flow of spray liquid. Thus, depending on the operating pressure, some are classified as 'extra coarse' nozzles while others would be in the 'coarse' category.

Cone nozzles have not been used on horizontal spray booms as the pattern across the width of the boom varies due to the overlapping of spray from adjacent nozzles, but they have been used on vertical booms known as 'drop-legs', mounted along a horizontal boom and on orchard sprayers. There are generally two types of cone nozzle: one has a disc with a central round orifice positioned next to a thicker disc, in which angled slots cause the liquid to spiral towards the orifice, so that with a range of different discs and cores a range of spray angles and outputs can be achieved. The second has the orifice with slots cut on the inside of the nozzle tip, adjacent to the orifice, with a flat plate screwed into the nozzle tip to close one side of the slots. This second type was useful when flow rates such as 0.2 l/min were used to spray cotton. The narrower-cone 60° nozzle (Conejet Y series) was preferred to the 80° (X series) nozzle, to project droplets at a slightly upward angle between layers of branches and to achieve deposition within the canopy and on the undersides of some leaves.

On some equipment, especially manually operated sprayers, there is now a requirement to fit a pressure control valve to ensure that the pressure, and thus the output, is constant irrespective of the variation in pressure from the pump. There are other special hydraulic nozzles, for example 'boomless' nozzles, to distribute spray where a boom is not practical, such as along the sides of roads.

Knapsack Hydraulic Sprayers

Despite the availability of knapsack sprayers in the 1950s, when DDT was being tried in Uganda on cotton (Jones, 1966), the recommendation was to use a double-acting slide pump or 'trombone' sprayer, which was laborious and led, no doubt, to an erratic distribution of spray. It was less expensive. However, the construction of knapsack sprayers, as with nozzles, adopted plastic technology. The earliest knapsack lever-operated plastic sprayer was the Policlair (Fig. 2.4). The spray tank, pressure cylinder and pump were all constructed with plastic materials, notably polypropylene, with a metal lance, pumping lever, crankshaft and some screws. Unfortunately, the mechanism to retain the diaphragm in the pump relied on a plastic ring screwed into the upper part of the pump. If the spray operator pumped energetically and exceeded the 0–30 psi pressure, increasing to 40 psi, the diaphragm was no longer held in position and the spray in the tank poured down over the operator's legs. There was also a problem with the lid, as it was not a screw fit and was difficult to open. Subsequently, a test procedure was developed that simulated operation of the sprayer on the operator's back and a redesign of the sprayer led to the development of the Cooper Pegler CP3 sprayer. This proved to be a very durable design and all manufacturers have adopted plastic spray tanks to minimize the cost of production, although some people still wanted the more traditional metal tanks.

In central Africa, a spray boom was attached to a pressure-retaining sprayer, the LeoColibri, marketed by Cooper Pegler in Rhodesia for tsetse spraying (Figs 2.5–2.6). These sprayers were extremely heavy so the 'tailboom' was attached to the back of the tank of a lever-operated knapsack sprayer (Tunstall *et al.*, 1961).

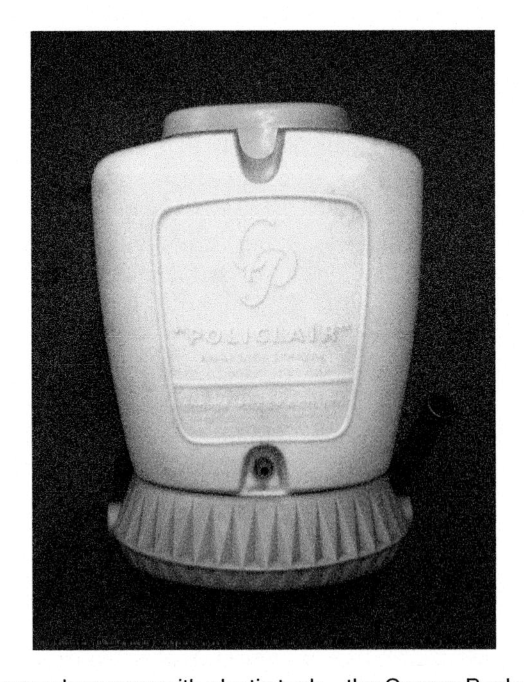

Fig. 2.4. First knapsack sprayer with plastic tank – the Cooper Pegler Policlair.

Fig. 2.5. Leo-Colibri sprayer being used to spray tsetse flies.

Fig. 2.6. Could a spray boom be used? Operators had great difficulty in walking at the same speed due to fear of meeting a snake in the cotton field.

Instead of the more usual brass tank, the Mysto K/GS/4 sprayer had a treated steel tank. The pump was operated by an overarm lever so that the operator's arm movement, generally, was above the plants. Using a 'tailboom', the volume of spray applied to cotton was increased with plant height – with cone nozzles mounted at different heights and directed slightly upwards to get spray projected into the crop canopy (Figs 2.7 a–e). This equipment enabled the very successful introduction to spraying cotton in Nyasaland and southern Rhodesia in the early 1960s using the insecticides supplied in sachets with the correct dose for a single tank load. Yields of seed cotton were more than doubled, but for many in the semi-arid areas, the volume of water that was needed prevented adoption of the technique. Later in this chapter the impact of subsequent research with ultra-low volume spraying is discussed.

In vector control, the application of DDT on wall surfaces to kill *Anopheles* mosquitoes, vector of malaria, required the use of a compression sprayer as the sprayer is pressurized before spraying and then the operator can direct the spray lance carefully at the wall surface. Initially, in the 1950s, some agricultural compression sprayers were tried, but to use them the pump had to be removed by the operator to fill the tank with spray liquid. This meant that insecticide on the pump could contaminate the operator while spray was being prepared, so a specification was published by the WHO requiring a compression sprayer with the pump separate from the tank lid. The first to meet this specification was the Hudson X-Pert sprayer, which was used extensively in the programme using DDT for

Fig. 2.7a. Testing the prototype vertical boom made with Dexion to position nozzles at different heights between the rows of cotton. (Note: table tennis balls on pins were used to collect samples of spray.)

Fig. 2.7b. An early 'tailboom' being tried on a farmer's cotton field.

Fig. 2.7c. Spraying tall cotton plants with a tailboom.

indoor residual spraying. Later similar sprayers were made available by other manufacturers, but it was not until 2016 that a plastic version met the WHO specification. One earlier attempt failed as the lid failed to retain pressure in the plastic tank. As the liquid is sprayed from a compression

Fig. 2.7d. Equipment for farmers in Malawi, 1960.

Fig. 2.7e. Introduction of sachets with sufficient insecticide for one sprayer tank load, 1960.

sprayer, the pressure falls, so a control flow valve must be fitted to the lance to provide a uniform spray at the nozzle (Figs 2.8a and 2.8b).

In agriculture, two other types of knapsack sprayer are marketed: a motorized knapsack sprayer is fitted with a two-stroke engine to drive a pump; or, more recently, a battery is fitted to drive an electrically powered pump. The motorized version is capable of very high pressures but is heavier than the electrically powered unit. The latter may have a heavy lead-acid battery but is more efficient with a lithium-ion battery. Some units are fitted with a multi-nozzle boom to treat a wide swath. These motorized units eliminate the drudgery of manual pumping but are more expensive, yet they have already become popular in parts of south-east Asia.

Arable Crops

All the hydraulic nozzle types discussed are designed to fit a nozzle body and cap that meets an international standard. The nozzle body should

Fig. 2.8a. Indoor residual spraying for mosquito control – compression sprayer with control flow valve.

Fig. 2.8b. Fan spray with control flow valve.

also have a filter fitted to avoid nozzles becoming clogged while spraying. The nozzle body should also have a cut-off valve so that once the spray has been turned off there is no pesticide dripping from the nozzles. This was first used on aircraft but is equally important on ground equipment to minimize spray on the margins of treated fields. For arable crops, spray booms that could deliver 50–500 l/ha were mounted on the tractor, although particulate suspension formulations then available were liable to cause blockages of nozzles at the lowest volumes applied (Figs 2.9 a, b). Boom width was still generally about 12 m, although much wider booms – from 21 m to 42 m – have become standard, especially where fields are relatively flat. The horizontal boom on tractor-mounted, trailed or

Fig. 2.9a. Tractor mounted sprayer in the UK, *c.*1950; no cab on tractor.

Fig. 2.9b. Trailed sprayer in the UK. (Photo courtesy of Househam.)

self-propelled sprayers is usually set so that the spray nozzles are 50 cm above the crop and direct the spray downwards. Now tractor cabs have global positioning systems (GPS) (Fig. 2.9c). Relatively few sprayers have been fitted with ‘drop legs’ – vertical booms to spray a row crop in much the same way as the tailboom referred to earlier. Apart from tractor-powered equipment there has been interest in animal-drawn sprayers, notably for use with oxen (Fig. 2.10), although a sprayer on a camel was tried in the Sudan.

Fig. 2.9c. Modern tractor cab controls with global positioning system (GPS). (Photo courtesy of Househam.)

Fig. 2.10. Ox-drawn sprayer, Rhodesia, 1965.

The inability to spray cereals infested with aphids with tractor equipment and insufficient aircraft for aerial spraying in the UK in 1976 led to a rapid adoption of 'tramlines' (unsown paths for the tractor wheels), developed by researchers examining the application of fertilizer at different times of crop growth. In some situations, usually with large areas to spray, boom height has been raised where faster speeds have been used, despite a potential increase in spray drift.

The filling of spray tanks was considered to be a health and safety issue if farmers needed to climb up onto the tank to pour a pesticide into the tank, so by the 1990s the sprayer had to be fitted with a low-level induction bowl (Fig. 2.11). This, then, incorporated a water supply to triple-rinse empty containers to reduce the hazards associated with disposal of contaminated containers. A separate small clean water tank was also required so that the operator could wash any deposit of pesticide off his gloves before removing them and for washing his hands. On some large farms, closed systems of pesticide transfer were adopted with return of containers for refilling.

Concern about drift from aircraft spraying cotton in Israel led to a return to ground equipment. One manufacturer added a sleeve across the length of the boom inflated by air from a fan so that a curtain of air was directed downwards to assist the spray droplets to penetrate the crop canopy (Fig. 2.12). This idea was soon copied by others (Taylor *et al.*, 1989; Cooke *et al.*, 1990), although to be effective, the crop has to have sufficient canopy to filter out the airborne droplets. This system was soon used by many farmers in contrast to earlier attempts to cover the boom or fit an aerofoil across the boom to direct spray downwards.

Fig. 2.11. Tractor sprayer with a low-level induction hopper to prepare sprays and a separate container for pesticide containers. Note the protective clothing worn by the spray operator. (Photo courtesy of Hardi.)

Fig. 2.12. Sprayer boom modified with air-inflated sleeve to project spray downwards into the crop. (Photo courtesy of Hardi.)

Depending on the volumes being applied and the pressure required, various pumps have been used on tractor-mounted and similar boom sprayers. Diaphragm pumps have proved reliable and popular, but in tropical countries, rotary vane pumps have been used and are effective.

Orchard crops

Potts (1958) had tried using an orchard duster to apply a concentrated spray, an oil-coated dust and a wet dust atomized into the airstream of a 4-in diameter discharge pipe, as early as 1928. He found that the air volume and velocity restricted the projection of either spray or dust to a height of 12 m. In 1937, a large air-blast high-volume sprayer was used in Florida to treat citrus. Initially it had four-bladed aircraft propellers to deliver a large volume of high-pressure air to form an arc where nozzles injected atomized liquid into the airstream (Rose, 1963). Later an axial fan replaced the aircraft propeller and the 'speed sprayer' became widely used, with several companies making different versions. The nozzle arrangement, air velocity and volume differed, with some having a V-shaped slot to project more air to the top of trees. As air velocity from these air-assisted sprayers reduces rapidly as the distance from the blowers increases, Potts (1958) pointed out the advantage of applying droplets in the 41–65-micron range for coverage at distances more than 15 m. He also pointed out that in North America alone during 1950–1957, an aggregate of over 150 million acres had been treated with concentrated sprays, increasing to as much as 30 million acres annually, much of which was applied aerially to protect forests in Canada and the USA. Dusts were used in the 1920s in the USA following the first demonstration of crop dusting with lead arsenate. The pilot John Macready took off from McCook field in a Curtiss JN-4 'Jenny' biplane on 3 August 1921 (Fig. 2.13). Aerial and ground application equipment had not been satisfactory because of poor deposition of dusts, their poor retention on foliage and other factors.

Labour costs for high-volume spraying was one of the factors that led to the Autoblast sprayer for orchards. In addition to the pump and spray tank, a powerful fan was driven by the power take-off to deliver 45,000 ft^3 of air/min to throw the spray through the orchard. One man could now treat 25 acres per day, so timing of application was better. Spray volumes were now reduced to one tenth of the previous spray volume – low-volume spraying was now possible.

In the 1950s, orchards were still sprayed in the UK with a manually directed lance and nozzle fitted to a high-pressure pump, albeit the tank and pump were now mounted on a tractor. An operator or a pair working together would spray each tree individually, so one man treated about 4 acres per day (Ordish, 1952). Spray volumes were generally greater than 2500 l/ha (250–500 gallons/acre). However, improved methods of application were now being sought, especially on high-value-per-acre crops.

Later studies in the UK led to the development of the Commandair sprayer, which projected 7.8 m^3/s to each side of the trailed sprayer at a

Fig. 2.13. First dust application from an aircraft, 1921. (Photo courtesy of Ohio Agricultural Experiment Station, Wooster, Ohio.)

velocity of 20.7 m/s (Hale, 1975). The nozzles were mounted in the airstream projected from the sides rather than in the more common design of mistblower with the axial fan at the end of the spray tank (Fig. 2.14a). A need for changes in sprayer design was influenced by health and safety regulations concerning using ladders to harvest fruit from apple trees and the amount of spray projected above the tree canopy, which could drift considerable distances downwind. The height of some tree crops was decreased by using dwarf rootstocks and pruning to achieve a trellis effect. On relatively flat ground, this enabled 'tunnel' sprayers to be used (Fig. 2.14b), which partially protected the spray from any wind and reduced drift (Planas *et al.*, 2002). More recently, an alternative development has been the air-assisted vertical booms that project the spray more laterally into the crop canopy without a tunnel, and enable more rows to be treated (Fig. 2.14c).

In Uzbekistan, mistblower-type sprayers were used on cotton, but the direction of the blower oscillated from side to side (Figs 2.15 a, b). Inevitably, spray directed against any wind was blown downwind and deposits varied significantly. This showed up very dramatically when the same sprayer was used to apply herbicide in a cereal crop. In Mozambique, they misused an air-blast sprayer by operating it in the middle of the day when thermal airflow lifted the spray up and away from the crop being treated (Fig. 2.15c).

Knapsack Mistblowers

During the early development of lindane to control mirids, a motorized fan was used in west Africa to blow spray from a knapsack sprayer with

Fig. 2.14a. Orchard spraying with axial fan on the back of the sprayer to project spray through the tree canopy (Photo courtesy of Long Ashton Research Station.)

Fig. 2.14b. 'Tunnel' sprayer to reduce downwind drift of spray in an orchard.

Fig. 2.14c. Modern unit with air-assisted vertical booms to treat multiple rows in an orchard.

Fig. 2.15a. Spraying cotton in Uzbekistan with oscillating blower.

Fig. 2.15b. Close-up of the Russian air blast sprayer.

Fig. 2.15c. Misuse of a tractor-mounted mistblower treating cotton in Mozambique at noon, *c.*1966.

a hydraulic nozzle higher into cocoa trees. Very rapidly, this evolved into the motorized knapsack mistblower with a two-stroke engine to drive a centrifugal fan, creating a high-velocity airstream that shattered the spray emitted through a restrictor into a mist spray. Initially, a version with the fan mounted to one side was considered (Fig. 2.16a), but soon the fan was positioned directly behind the operator's back with the spray tank high up to deliver the spray by gravity (Fig.2.16b).

The fuel tank was initially mounted high on the knapsack frame, but later, for safety reasons, it was mounted below the engine. Later, the spray tank was slightly pressurized and on some models a pump was driven by the motor to ensure more uniform spraying irrespective of the height of the spray nozzle. Various nozzles have been tried, including a simple tube, but a more uniform spray is produced if the spray liquid is fed onto a flat plate or cone to form a thin sheet shattered by the airflow. On some machines a rotary nozzle has been used. Apart from spraying cocoa, knapsack mistblowers have been used on many other tree crops, such as mango, and when used to spray horizontally, they have applied insecticides on cotton and other field crops. Their use on cotton by spray teams in one country resulted in most sprayers being returned to the workshop for repair after relatively little use. This was because operators preferred to keep the two-stroke engine idling while walking from one farmer's crop to the next, ignoring the fact that the engines were not designed to operate at idle speed for long periods. Similar problems were reported when trying to use mistblowers in vector control, where they have been used for indoor residual spraying to treat houses quickly. They are also used to treat vegetation around houses to create a barrier to deter mosquitoes reaching houses. Adapted as a granule applicator, they can be used to disperse larvicides over a large area.

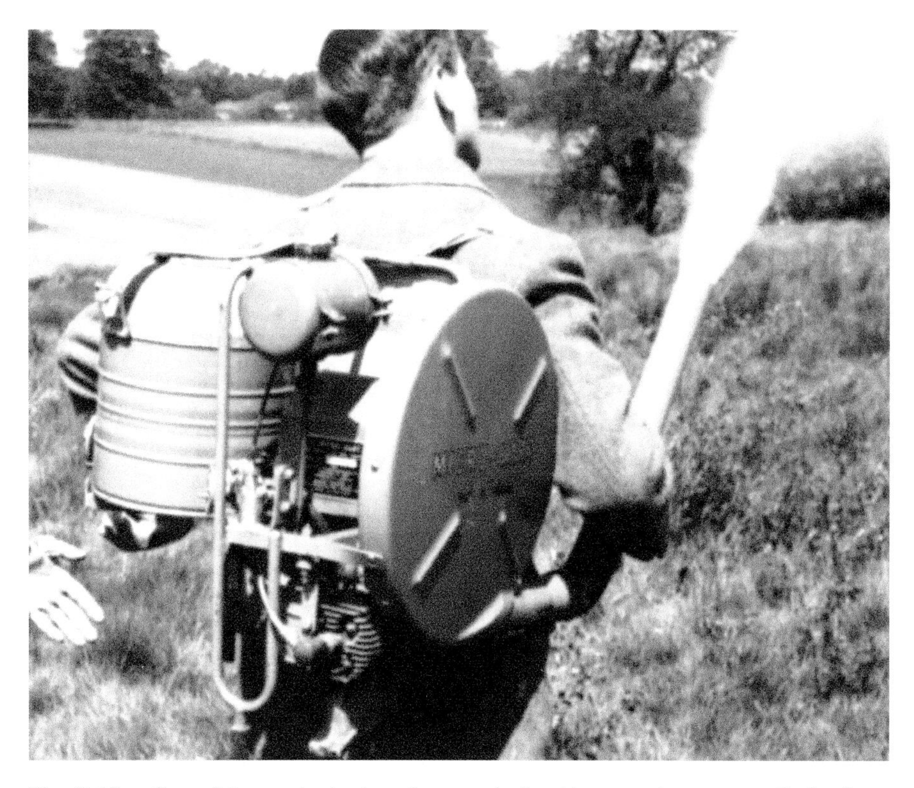

Fig. 2.16a. One of the early designs for a motorized knapsack sprayer with the fan on the side, 1957.

Fig. 2.16b. Motorized knapsack mistblower (Stihl), showing fan now mounted across the back of the operator.

Ultra-low-volume Sprays

The need to spray against desert locusts in Africa immediately after World War II resulted in efforts to spray minimal volumes over vast areas. Much was done by the Anti-Locust Research Centre, set up by the Colonial Office in 1945, working with the Colonial Insecticides Research Unit (now Tropical Pesticides Research Institute (TPRI)) at Arusha, Tanzania. The studies on application led to two important developments. One development was the exhaust gas nozzle sprayer (ENS) fitted to a Land Rover, which used the engine exhaust to atomize the spray, which was projected upwards to allow dispersal of droplets downwind over a wide swath. A 5–10% oil formulation of dieldrin was effective when applied with as little as 0.3 l/ha. The technique was known as vegetation baiting as it deposited the insecticide on the sparse vegetation being eaten by the locust nymphs ('hoppers'). Much later, the ENS was replaced by a sprayer with multiple spinning discs mounted on a mast (Ulvamast) to provide a more uniform droplet size.

The other development was a rotary nozzle, the Miconair, fitted to aircraft. Edward Bals, who had been witnessing the mud sprayed on his windscreen from the wheels of the vehicle in front of him, decided to experiment with a rotary nozzle, and in 1954 he made an atomizer with a large cylinder of metal gauze, which was developed by Britten-Norman Ltd as the Micronair A100 atomizer, two of which were initially mounted on a Tiger Moth and driven by a V-belt from a windmill. Later this was further developed, and in 1956 a larger cage atomizer was driven directly from a windmill – the Micronair A1000. The blades of the windmill could be adjusted to change the number of revolutions/min of the atomizers and thus the droplet size. Further development led to the AU3000. Apart from locust control, applying usually 1 l/ha, the aerial unit was used to spray cotton in the Sudan Gezira, as well as other crops such as bananas. The latest versions are the AU4000 and AU5000 rotary atomizers on aircraft and AU7000 and AU6539 on helicopters.

Edward Bals later designed a large disc mounted directly to an engine making the Micron tomato sprayer, but miniaturizing this resulted in the hand-carried, battery-operated Turbair X (Fig. 2.17a), and later the Micron ULVA 16 (Fig. 2.17b), both of which successfully applied up to 3 l/ha of insecticide in an oil carrier rather than diluted in water. On cotton, ULV sprays resulted in similar yields obtained with a knapsack sprayer and 'tail-boom' applying up to 200 l/ha. Further improvements with the motor and a smaller disc with grooves to the peripheral teeth resulted in the ULVA + (Fig. 2.17c), which was extensively used on cotton in francophone west Africa, changing from ULV sprays to the application of water-based formulations of insecticide applied at 10 l/ha from 1995. This was due to industry discontinuing support for the oil-based spray despite the distinct advantage of persistence of oil droplets on leaf surfaces following rainfall. Carrying the spinning disc downwind of the operator, droplets of 70–100 microns disperse over a relatively narrow swath and the operator is less exposed to insecticide than when nozzles on a lance are held in front of the operator.

Fig. 2.17a. Early ULV sprayer for treating cotton crops in Africa.

Fig. 2.17b. Initial trials with ULV spraying in Malawi, 1969.

Noting the success of ULV spraying, Coffee (1979) developed a means of atomizing an oil-based liquid using relatively little power from small torch batteries to a generator that delivered 25 Kv to a semi-conducting liquid as it emerged by gravity though a narrow annular slit – the Electrodyn sprayer (Figs 2.18 a, b).

Fig. 2.17c. Adoption of ULV spraying of cotton in Cameroon, *c.*1988.

Fig. 2.18a. Prototype Electrodyn sprayer in Sudan.

The liquid broke into ligaments and then droplets with a narrow droplet size range. A ULV spray was effective on cotton in semi-arid areas, but lush growth was not as effectively treated, as charged droplets are deposited on the nearest earthed objects, i.e. the leaves nearest to the spray droplets. Thus, penetration into a crop canopy is not so effective without air movement dispersing droplets within the foliage (Fig. 2.19). Other electrostatic sprayers have been developed, but none has been very successful commercially. In the UK, a system of charging droplets on a tractor sprayer boom was marketed, but farmers were concerned that the spray deposit was confined to the upper part of a wheat crop and failed

Fig. 2.18b. Electrodyn 'Bozzle' being used in Brazil.

Fig. 2.19. Electrostatic spray showing movement of droplets around a single plant. (Photo courtesy of Silsoe.)

to penetrate to the stems and lowest section of leaves to control diseases (Fig. 2.20). In the USA, studies by Law (Fig. 2.21) involved charging small droplets created within an airstream, but their use has not been adopted on a large scale. Later, using droplets around 100 μm in an airstream, a battery operated portable electrostatic sprayer has been developed primarily for spraying disinfectants in hospitals; it could also be used for a rapid insecticide spray.

For weed management, spinning discs operated at lower speeds to create large droplets so that a herbicide spray is deposited downwards on the weeds. The 'Herbi' and 'Handy' (Fig. 2.22) sprayers were designed to remove the drudgery of manual weeding of tropical crops. On some versions, the disc is protected by a shroud to prevent any droplets travelling further to deposit spray on the crop. Other versions of the spinning

Fig. 2.20. Electrostatic system with hydraulic fan nozzle.

disc technology have been developed including an air-assisted version (Ulvafan) for applications in glasshouses.

Apart from locust work, TPRI did considerable research on aerial applications, notably to control tsetse flies. Early work in Uganda showed that significant control was obtained only when using coarse aerosols from feeding the insecticide into a modified engine exhaust system (Hocking and Yeo, 1953). Later laboratory studies indicated droplets around 30 µm would be effectively filtered by sedentary flies (Hadaway and Barlow, 1965). Vast areas with tsetse flies were treated in southern Africa with aerial sprays (Allsopp, 1984), but in the Okavango Swamp area of Botswana, an alternative technique was tried, following the ban on applying endosulfan. This involved soaking a blue-coloured material in insecticide, so that when dry, the cloth could be hung together with a phial containing an attractant. Tsetse flies landing on the treated cloth received a lethal dose. Apart from some traps being knocked over by buffalo, heavy rains in Botswana prevented access to the traps, so they returned to aerial spraying (Allsopp and Phillemon-Motsu, 2002). This is now far more accurate with global positioning systems to ensure even application over large blocks and to control exactly where spray is applied. Kgori *et al.* (2006) reported that application of deltamethrin, applied at 0.26–0.3 g/ha sequential aerosol treatment (SAT) five times to control *Glossina morsitans* was highly successful. A similar night-spraying SAT programme was successfully conducted in Ghana against *Glossina palpalis gambiensis* and *G. tachinoides* combined with ground-spraying very dense forest areas and treating cattle (ITC), as well as deploying insecticide treated targets (ITT) (Adam *et al.*, 2013). Sequential treatments are essential as the dose only affects the adult tsetse flies at the time of application and has no impact on immature

Fig. 2.21. Electrostatic spraying: Ed Law with his tractor-mounted ES sprayer.

stages, which can emerge later. The gap between the treatments depends on temperature and other factors affecting their development.

Aerial Spraying

Zimmerman proposed the use of aircraft to control forest pests as early as 1911 (Maan, 1961). Apart from forests being treated with lead arsenate to

Fig. 2.22. Using 'Handy' sprayer for weed control. (Photo courtesy of Micron.)

control catalpa sphinx (*Ceratomia catalpa*) (Houser, 1922) cotton crops were also dusted with arsenate insecticides.

One of the earliest aircraft to be used in the 1920s was a World War I 'Jenny' biplane, used to spread dusts to control cotton boll weevils. Other early biplanes were the Grumman Ag-Cat and the Russian Antanov An-2, which was still being used in the 1980s (Fig. 2.23a). A Tiger Moth was also converted for aerial spraying. The first aeroplane specifically designed for applying pesticides was the Ag-1, developed in Texas in 1949–1950, from which the specifications for aircraft applying pesticides were developed (Quantick, 1985). Following Ag-1, a number of spray planes have been used to apply pesticides as well as to fight fires. These include the Piper PA-25 Pawnee (from 1959), the Cessna Ag-Wagon (from 1960), Air Tractor, Thrush Commander, Brazilian Embraer EB 203 Ipanema and Polish Dromader M-18. Helicopters used for spraying included the Sikorsky R4,

Fig. 2.23a. Antanov biplane used to spray cotton in Egypt, 1980s.

Hiller 12E, Hughes 500 and Bell Jet Ranger. Larger aircraft have included the DC3 Dakota, DC7 and C130H, especially when controlling mosquitoes over extensive areas. In a few places, insecticides have been applied using micro-light aircraft. Following trials with autogyros in the 1930s (Potts, 1939), Dr W.E. Ripper used the Sikorsky R4 helicopter probably as early as 1944 to determine the effect of rotor downwash on the distribution of spray droplets. In practice, due to the economics of operating helicopters, the flying speed results in a similar downwind distribution as a fixed-wing aircraft (Fig.2.23b).

Spray aircraft were initially fitted with hydraulic nozzles on a boom mounted along the trailing edge of the wing, avoiding the wing-tip area due to wing-tip vortices that can lift spray upwards. Later rotary atomizers were more widely used, initially for locust control and field crop spraying (Fig. 2.23 c, d). Public perception is that there is significantly more downwind drift, and now the EU has banned aerial spraying, unless a country requests a derogation for a specific reason. Thus some aerial applications have continued over forests and wet fields, unsuitable for tractors, and in a few instances to control vectors of human diseases.

Spray volume application rates from aircraft were always less than those used with ground equipment but are kept to a minimum, especially with the development of unmanned aerial vehicles (UAV), also referred to as remotely piloted aircraft (RPA) or drones. UAVs have been used in Japan since the 1990s to treat rice fields. There are now over 2000 Yamaha RMax helicopters in Japan. The RMax, controlled by a ground operator, is powered by a two-stroke, two-cylinder 2.4-litre engine, can lift a payload of 28 kg and spray at about 24 km/h. Further development of UAVs is likely to provide a more robust platform with GPS/GIS control, which

Fig. 2.23b. Helicopter spraying cotton in Rhodesia, 1962.

Fig. 2.23c. Spraying locusts. (Photo courtesy of FAO.)

Fig. 2.23d. Aerial spraying – checking swath width with dyed spray.

can apply ULV pesticide sprays close to the crop canopy at a higher speed than that achieved with large ground equipment, thus reducing the impact of soil compaction along 'tramlines' through the fields. Related to less soil compaction, ploughing could be minimized with an increase in 'no-till' technology to control weeds, involving UAVs to apply herbicides.

Agriculture

Large areas of cotton, bananas, potatoes, sugar cane, rice and other crops have been sprayed aerially in many countries, especially in areas where soil conditions are not suitable for ground equipment or where rapid treatment is required. Fungicide application on bananas to protect young leaves from disease was much more effective than attempting to deliver spray from ground equipment. In the USA, aerial spraying has been extensive, but concern about spray drift led to a Spray Drift Task Force that ultimately introduced a scheme based on the BCPC spray classification of nozzles (Hewitt, 2008).

Forestry

Various application methods are needed in forestry to cover different situations from seedlings to mature trees, from hand-operated equipment to aerial spraying, using a range of aircraft, from large, fixed-wing planes to small helicopters. In contrast to agriculture, there is a greater need to minimize adverse effects on non-target species, so equipment has to handle biopesticides, such as *Bacillus thuringiensis* aimed at certain lepidopteran pests. Vast areas of forests in North America have been sprayed aerially to protect trees, particularly for the control of defoliation of spruce and fir trees by the spruce budworm. Elsewhere, sudden outbreaks of various pests, including pine beauty moth and nun moth, and diseases have been treated using aircraft. Since the 1970s considerable research has been conducted to improve the spray technology to optimize dose transfer. Now with the aid of modelling and other factors there are manuals to guide aerial spray programmes.

Vector control

Apart from controlling tsetse flies, aircraft have been used to apply larvicides to control mosquitoes transmitting malaria and other diseases, and blackflies (*Simulium* spp.), the vector of onchocerciasis, causing river blindness, notably in a 20-year programme covering nine countries in the Sahel area of west Africa to break the transmission of the parasite. In the Onchocerciasis Control Programme (OCP), 14,000 km of river needed surveillance and treatment so helicopters (Bell Jet Ranger) and fixed-wing aircraft (e.g. Pilatus Porter) were used with a rapid release system to discharge a dose related to river flow, just upstream of rapids with 'white water' where the *Simulium* larvae are found (Parker, 1975). The objective was for the insecticide to spread across the river so that the width of the rapids had a specific dose for a short period. Although the programme was successful, the cost was high.

Although malaria was eliminated from the USA, the main concern was that people dislike the nuisance of being bitten by mosquitoes, so aircraft have been used both to apply adulticides as a space spray with very small droplets or larvicides with rather larger droplets or as granules. In the USA, aerial sprays against mosquitoes had been with hydraulic nozzles, but a trial with special high-pressure nozzles showed that when most of the spray was in very small droplets, mortality of mosquitoes was much higher and over an area much further downwind than previously. An alternative to aerial spraying has been to apply a fog from a vehicle, supplanted with manually carried equipment where there is no access for vehicles.

In Africa, apart from a trial in Tanzania (Yeo and Wilson, 1958), aerial spraying has not been adopted to control mosquitoes. The main use of insecticides against both *Anopheles* and *Aedes* spp. has been with indoor residual sprays (IRS) or distribution of insecticide-treated bed nets (ITNs). However, neither of these techniques reduces the risk of disease transmission when people are bitten outside their houses.

Another aspect of pesticides and aircraft is the process of aircraft disinsection, aimed at preventing insects being transported from one country to another. A standard procedure on passenger aircraft has been to apply a space spray (aerosol) in the cabin after boarding, but prior to the aircraft taking off. In some cases this may be a pre-embarkation treatment or it may be delayed until the plane is about to descend to the next airport. Luggage and flight compartments should also be treated with a residual spray, and to keep cockroaches and other pests in the food areas a gel or residual spray has been applied primarily to floor areas and behind galley panels. Toilet areas are also treated.

Space Treatment

Historically, space treatments were initiated to treat the inside of buildings and glasshouses. This was with a thermal fog, with which the insecticide diluted in a liquid, such as kerosene, is injected into a stream of hot air, with a temperature of 500°C. The liquid is vaporized, but condenses as very small droplets, usually <20 μm, which remain airborne. The technique was used mainly in warehouses and timed to coincide with the flight times of insect pests of stored products. Thermal fogs were also used in vector control aiming to kill adult mosquitoes in flight using a low dose, non-persistent insecticide. The white cloud produced by a thermal fog is very obvious. This is appreciated by the local population, which recognizes control measures are being implemented. However, dispersal of large quantities of diluent is not so acceptable in the environment, so application is now usually with cold fogs.

An insecticide formulation with adjuvants to reduce evaporation of the spray droplets is applied using a vortex of air to produce the very small droplets. Although less spectacular, the impact of cold fogs applied using ultra-low volumes of spray liquid is similar to a thermal fog. Both techniques can be used to control pests inside glasshouses and warehouses, but also are sometimes used inside dwellings to control mosquitoes. In warehouses, electrically operated cold foggers can be preset to function at a specific time when the insect pest is most active. As explained earlier, sequential treatments are required, as only one part of the pest life cycle is affected by a space treatment.

Seed Treatment

Treating seeds before sowing dates back to Roman times, when seed was mixed with ashes and other materials to protect young seedlings from soil-borne fungi or other pests. This increased in the 19th century with the use of 'bluestone' (copper sulfate) and organo-mercurial compounds, joined much later by some of the new pesticides, such as thimet in the 1950s. Gamma BHC was a new seed treatment introduced in the 1940s as an effective wireworm control and later replaced by a cyclodiene such as dieldrin and combined with a fungicide.

Later, the use of a systemic insecticide as a seed treatment was welcomed as it provided good control of foliar sucking pests, particularly during the early stage of plant growth. This was a technique adopted with the development of neonicotinoid insecticides as the seed treatment eliminated the need to spray small plants. Unfortunately, early use of some neonicotinoid-treated seed resulted in high bee mortality. This was due to inadequate retention of the dust on seed coupled with old-style planters, which exhausted air upwards into the atmosphere. Thus insecticide dislodged from seeds was blown upwards into the air and subsequent downwind drift adversely affected bees. Improved formulation of the insecticide gave much higher retention on seed, while retrofitting seeders, to project any dust downwards into the soil, minimized downwind drift of dust. In Europe, there was a moratorium on sowing treated seeds; however, farmers prevented from using treated seed began spraying other insecticides on their crops.

Wiping rather than Spraying

Sometimes there are localized patches of weeds or a weed that is taller than the crop. Various designs of a 'wet' surface fed from a reservoir of liquid have been made to wipe the weeds with a herbicide (Fig. 2.24). It is crucial that for spot treatments, such a weed-wiper applicator is designed to avoid herbicide dripping onto a crop. Essentially, a surface kept moist with a herbicide is used to touch weeds and leave sufficient active herbicide on the weed. The technique has been most effective with a herbicide such as glyphosate, which is redistributed down the plant. It is also effective where weeds are taller than their surroundings. Weed wipers vary in width from hand-carried units to those mounted across the width of a tractor boom. They have been used mostly in pastures and other grassland, low-level crops and ecologically sensitive areas. No drift is created as the transfer is only by direct contact with the weed, and the wiper should be designed to avoid herbicide dripping from the wiper. Apart from glyphosate, several other herbicides have been applied using weed wipers, including metsulfuron, clopyralid, triclopyr and picloram (Harrington and Ghanizadeh, 2017).

Personal Protective Equipment (PPE)

Pesticides vary in their toxicity for humans, so the WHO created a classification system that rated pesticides from extremely toxic (Ia) through Ib (highly hazardous), II (moderately hazardous), III (slightly hazardous) and U (unclassified). In some countries the extremely hazardous pesticides were permitted only if formulated as granules for treating the soil, but with increasing awareness of the problems of ill health of users, if not properly protected, registration of the highly hazardous pesticides is now being withdrawn in many countries. The preparation of sprays was recognized as the main activity where farmers would be most exposed to the pesticides. It was generally considered that for most pesticides in the

Fig. 2.24. Weed wiper being used in rubber, Indonesia.

1950s, once they were diluted, there was less risk to the spray operator. Nevertheless, those applying high volumes of spray would wear a mackintosh, hat and wellington boots to avoid being wetted by the spray.

In the UK, spray operators have had to attend training and obtain a certificate of competence for the relevant equipment to be used since the 1980s, to comply with the Food and Environment Protection Act 1985. BCPC published a number of handbook guides on boom sprayers, nozzles and other equipment to help farmers and ensure they followed the Code of Practice for the Safe Use of Pesticides on Farms and Holdings. Now, those born before 31 December 1964 or with a grandfather's exemption have to comply with regulations under the EU Sustainable Use of Pesticides Directive (SUD). Thus, most obtain the certificate for applying sprays with a tractor-mounted, drawn unit or a self-propelled sprayer. Others may get it for a manually operated sprayer, mostly using a knapsack sprayer, while others using pesticides in stores require training in using fogs. This training includes the use of the appropriate personal protective equipment. The National Register of Spray Operators (NRSO) has over 20,000 members who are required to participate in continuing professional development (CPD).

The EU Sustainable Use of Pesticides Directive also requires sprayers to be tested regularly to ensure they do not leak, and that all nozzles are applying the correct spray volume. In the UK, the National Sprayer Testing Scheme (NSTS) is responsible for tests. New equipment is tested once it is five years old. Smaller equipment, notably knapsack sprayers, do not require a formal test but should be inspected regularly following a checklist, and a record kept.

Unfortunately, many countries, especially in the tropics, lack the trained staff to implement adequate training and testing of equipment, so their farmers remain exposed to the risks associated with applying pesticides. At the same time, availability of PPE is often poor, or considered too expensive by the majority of small-scale farmers who rely on manually carried equipment. The use of lever-operated knapsack sprayers inevitably results in the operator being exposed to the pesticides. The formulation is often provided in litre or larger quantities, so a small amount has to be measured out for each sprayer load. When the idea of sachets containing the correct quantity for a tank load was first suggested, it was considered an expensive method, although it prevented spillage especially with wettable powders, as an opened packet of white powder could be mistaken for flour in many situations and where farmers kept pesticides in their house. Farmers inevitably walk with the spray lance in front of their body, so walk towards the spray, with treated foliage touching their legs. As mentioned earlier, where sachets were used with a tailboom behind the operator, it was possible to minimize operator exposure, and farmers were advised to wear a long-sleeved shirt and long trousers to reduce the area of skin exposed to the spray.

Environmental Protection

Using pesticides on farms raised concerns about spray drift from fields reaching water courses. However, the most serious contamination of water was shown to be due to accidental spillage or washing down of equipment on hard surfaces, so that water containing pesticides entered drains. Other sources were due to rainfall washing spray deposits from foliage and the subsequent 'run-off' entering ditches, going directly into streams alongside fields or percolating through the soil to reach ground water. Spray drift was the cause of the least exposure of water to pesticides; nevertheless, spray drift was a concern, due to possible adverse effects on vegetation downwind and the health of bystanders, residents and wildlife (Gilbert, 1995). The studies, originally in the UK, indicated that most droplets sedimented within a short distance from the crop edge, but very small droplets could travel much longer distances, especially if a thermal air current took the droplets upwards initially; thus the concept of using 'buffer zones' or no-spray areas downwind of a treated field. The importance of a buffer strip was recognized but interpreted differently depending on factors such as the extent of canals, as in Holland, or the size of fields. In the UK, the decision was to have a Local Environmental Risk Assessment of Pesticides (LERAP) with an unsprayed buffer of 1–5 m from the top of the bank of a ditch, depending on the pesticide, the spray nozzles/equipment used and dosage being applied (Gilbert, 2000). Subsequently, wider buffer zones were introduced for certain pesticides to retain registration of the pesticide, although farmers preferred the narrow buffer to avoid losing a large portion of the crop area within a field. Having vegetation in the buffer strip and hedge alongside a field collected most of the spray drift, but with integrated pest management, exposing the hedge to spray could have an adverse impact on some natural enemies. Farmers have tended to use one type of nozzle to treat a whole field, but with GPS widely used now it would be possible to apply a coarse spray over the last swath downwind or omit spraying close to the field's edge. This could result in the need to subsequently manage the 'unsprayed' strip so that persistent weeds cannot survive and spread seeds.

Apart from spillages, the washing out of sprayers after a treatment was one source of contamination of water. One way to reduce this problem was to wash the sprayer out in the field that has been treated and not back in the farmyard on a hard surface. Another technique developed in Scandinavia has been to create a biobed – essentially a pit lined with plastic containing peat and straw – to filter the washings from the sprayer and allow bacteria to break down the diluted pesticides.

Formulation

An important part of successful pesticide application is how the 'active ingredient', whether chemical or biological, is available to the user. Sprays have predominately been applied with water as a diluent, so the initial products were an emulsifiable concentrate (EC) or wettable powder (WP) formulation.

The EC was produced by dissolving the active ingredient in a suitable solvent to which an emulsifier was added, so that when mixed with water it formed an oil-in-water suspension of oil globules dispersed in the water. The spray was a white colour and droplets on leaves would spread over the surface, and some of the active ingredient would be absorbed into the plant. The general perception of users prior to 1960 was to use a large volume of water to wet plants, although using high volumes meant that once the foliage was wetted, excess liquid dripped down to the soil's surface. Using a large volume of water also diluted the surfactant activity of the formulation, so many hydrophobic leaf surfaces remained very difficult to wet.

The major trends in formulation have been to improve safety to the operator and allow much less water to be used, so spray volumes have deceased from 1000 l/ha or higher to around 200 l/ha, with many farmers now using <150 l/ha, especially with high tractor speeds. Concern about the risk of inhaling the very small particles in a WP led to the development of wettable granules (WG), which are too large to inhale. These are now preferred and are designed to disintegrate on contact with water and form a suspension. They have one disadvantage of measuring, when less than one container quantity is needed.

The health concerns of solvents in EC has resulted in the use of suspension concentrates (SC). Essentially, the pesticide in very small particles is mixed in a liquid that is designed to keep the particles in suspension, which can be measured out in a similar way to an EC. Early attempts to make suspension concentrates were unsuccessful as the particles gravitated downwards and formed sludge in the container, which was difficult to re-suspend. Improved persistence of more volatile actives can be achieved by using encapsulation, developed initially as a means of spraying insect pheromones, but now used for more volatile chemicals, especially as refinements of the microcapsule can allow different release rates from spray deposits and, where necessary, increase persistence of activity. Particulate formulations are often more effective in killing insect pests that are active on the surface of foliage and also on wall surfaces on which mosquitoes may rest.

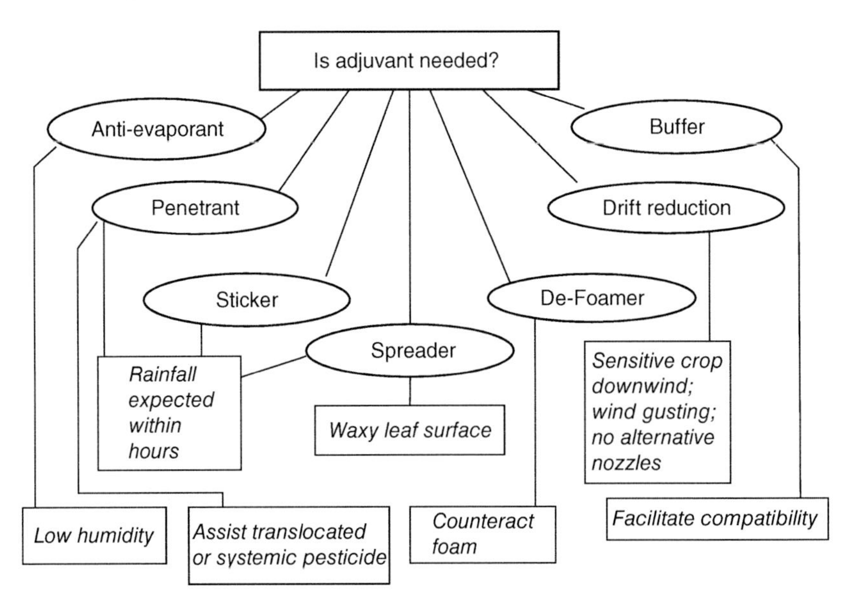

Fig. 2.25. Uses for adjuvants mixed with sprays.

Manufacturers feel that the farmer need not add anything to a spray, but there are many products referred to as adjuvants that are commercially available, which may provide some additional benefit, such as less spray drift, increased protection from rain or enhanced penetration into plants (Fig. 2.25).

Packaging

The size, shape and type of packaging of pesticides is an important issue as the user has potential exposure to the concentrated formulation of the active ingredient before diluting, usually in water when preparing sprays. Pesticide containers, their design and their handling are subject to stringent regulations including those from the EU, the Water Directive, health and safety requirements and the Waste Directive.

Detection of pesticides in water was traced in some situations due to careless discarding of the foil seals on containers that were subsequently washed by heavy rain into nearby drains. Similarly, splashes and spillage of pesticides can inadvertently contaminate local water. Triple rinsing of containers is now advised to minimize the amount remaining in a container and adding the rinsate into the sprayer reduces the residue in empty containers. Farmers need a system of storage of their pesticide and subsequent management of the containers.

Pesticides have been sold in various types of containers. One of the earliest designs was a tin can with a flat top surrounded by a small flange, which had an opening to one side of the top. Pouring of liquid from one of these can often cause glugging and pesticide remaining on the top of the can. This residual liquid was one of the problems of sheep farmers being exposed to organophosphate insecticides such as diazinon. Overseas pesticides were often delivered in much larger containers, usually 200 l, and local repackaging was often carried out.

For the farmer, pesticides are now marketed mostly in plastic containers that have a funnel shape towards the top as they are easier to rinse, and the sloping outside surface does not collect liquid. The opening needs to be wide enough to avoid glugging. There should be a handle on larger containers, and this should be designed so that the inside of the handle is not exposed to the pesticide.

The size of containers is best determined by the estimated quantity of pesticide needed, to avoid having too much pesticide at the end of the season and having to dispose of too many small empty containers. For the small-scale farmer, sachets with a quantity suitable for a knapsack sprayer are ideal. Some are marketed inside a foil protection, so that the whole sachet of a water-soluble plastic is added to the sprayer as this minimizes exposing the user's fingers to chemical while measuring out small quantities from a larger container (Fig. 2.26) (see also Chapter 9). On large farms, 2-, 3-, 5- and 20-litre, or larger, containers are usually available. Stores need to be designed so that stocks of chemicals are used in sequence to avoid accumulating an obsolete stock requiring special arrangements for disposal, a major problem in many

Fig. 2.26. Sachets of pesticide.

countries where the pesticide may arrive after the pest is no longer a problem or, as in the situation with locusts, may have migrated elsewhere.

Empty containers after triple-rinsing need to be taken as hazardous waste for recycling. This is complicated as plastic containers made from different polymers, e.g. polyethylene terephthalate (PET) and high-density polyethylene (HDPE), need to be kept separate. Some countries have developed good systems of recycling containers but in many countries disposal of containers has remained a problem.

To eliminate exposure of the operator to pesticides during preparation of sprays, closed transfer systems have been developed. Progress has been slow as industry has been reluctant to support multi-trip containers. On tractor equipment, using systems such as the FasTrans connection to the sprayer, the operator has had to remove the container cap and any foil seal before attaching it to the closed transfer unit on the sprayer and transferring the contents to the sprayer tank. Looking at other industries, using closed transfer systems has led to new ideas on connecting the container to the sprayer without the operator having to open the specially designed container.

References

Adam, Y., Cecchi, G., Kgori, P.M., Marcotty, T., Mahama, C.I. *et al.* (2013) The sequential aerosol technique: a major component in an integrated strategy of intervention against riverine tsetse in Ghana. *PLOS Neglected Tropical Diseases* 7(3), e2135.

Allsopp, R. (1984) Control of tsetse flies (Diptera: Glossinidae) using insecticides: a review and future prospects. *Bulletin of Entomological Research* 74, 1–23.

Allsopp, R. and Phillemon-Motsu, T.K. (2002) Tsetse control in Botswana – a reversal in strategy. *Pesticide Outlook* 13, 73–78.
Brown, A.W.A. (1951) *Insect Control by Chemicals.* Wiley, New York.
Brunskill, R.T. (1956) Factors affecting the retention of spray droplets on leaves. *Proceedings of the 3rd British Weed Control Conference* 2, 593–603.
Coffee, R.A. (1979) Electrodynamic energy: a new approach to pesticide application. *Proceedings of the British Crop Protection Council Conference – Pests and Diseases,* 777–789. BCPC, Farnham, UK.
Cooke, B.K., Hislop, E.C., Herrington, P.J., Western, N.M. and Humpherson-Jones, F. (1990) Air-assisted spraying of arable crops in relation to deposition, drift and pesticide performance. *Crop Protection* 9, 303–311.
Fraser, R.P. (1956) The mechanics of producing sprays of different characteristics. *Proceedings of the Second International Conference on Plant Protection.* Butterworth Scientific, London, pp. 237–277.
Gilbert, A.J. (1995) Analysis of exposure to pesticides applied in a regulated environment. In: Best, G.A. and Ruthven, A.D. (eds) *Pesticides – Developments, Impacts and Controls.* Royal Society of Chemistry, Cambridge, UK.
Gilbert, A.J. (2000) Local environmental risk assessment of pesticides (LERAP) in the UK. *Aspects of Applied Biology* 57, 83–90.
Hadaway, A.B. and Barlow, F. (1965) Studies on the deposition of oil drops. *Annals of Applied Biology* 55, 267–274.
Hale, O.D. (1975) Development of a wind tunnel model technique for orchard spray application research. *Journal of Agricultural Engineering Research* 20, 303–317.
Harrington, K.C. and Ghanizadeh, H. (2017) Herbicide application using wiper applicators: a review. *Crop Protection* 102, 56–62.
Hewitt, A.J. (2008) Droplet size spectra classification categories in aerial application scenarios. *Crop Protection* 27, 1284–1288.
Hocking, K.S. and Yeo, D. (1953) Aircraft application of insecticides in West Africa. 1. Preliminary experiments in areas supporting populations of the tsetse fly (*Glossina palpalis*). *Bulletin of Entomological Research* 44, 589–600.
Houser, J.S. (1922) The airplane in Catalpa shinx control. *Ohio Agricultural Experiment Station Bulletin* 7, 126–136.
Jensen, P.K., Jorgensen, L.N. and Kirknel, E. (2001) Biological efficacy of herbicides and fungicides applied with low-drift and twin-fluid nozzles. *Crop Protection* 20, 57–64.
Jones, T.R. (1966) Comparison of hand-operated machines for cotton pest control in Uganda. Part I – Description of machines under test. *East African Agricultural and Forestry Journal* 31, 409–415.
Kgori, P.M., Modo, S. and Torr, S.J. (2006) The use of aerial spraying to eliminate tsetse from the Okavango Delta of Botswana. *Acta Tropica* 99, 184–199.
Lodeman, E.G. (1896) *The Spraying of Plants.* Macmillan, London.
Maan, W.J. (1961) Fifty years of agricultural aviation. *Agricultural Aviation* 3, 77–81.
Matthews, G.A., Bateman, R. and Miller, P. (2014) *Pesticide Application Methods* (4th edn). Wiley-Blackwell, Oxford, UK.
May, K.R. (1950) The measurement of airborne droplets by the magnesium oxide method. *Journal of Scientific Instruments* 27, 128–130.
Miller, P. (2014) Spray drift. In: Matthews, G.A., Bateman, R. and Miller, P. (eds) *Pesticide Application Methods* (4th edn). Wiley-Blackwell, Oxford, UK.
Ordish, G. (1952) *Untaken Harvest.* Constable, London.
Parker, J.D. (1975) The use of aircraft in the WHO Onchocerciasis control programme. *Proceedings of the Fifth International Agricultural Aviation Congress,* 127–136.
Planas, S., Solanelles, F. and Fillat, A. (2002) Assessment of recycling tunnel sprayers in Mediterranean vineyards and apple orchards. *Biosystems Engineering* 82, 45–52.

Potts, S.F. (1939) Spraying woodlands with an autogyro for control of gypsy moth. *Journal of Economic Entomology* 32, 381–387.

Potts, S.F. (1958) *Concentrated Spray Equipment, Mixtures and Application Methods.* Dorland Books, New Jersey, USA.

Quantick, H.R. (1985) *Aviation in Crop Protection, Pollution and Insect Control.* Collins, London.

Rose, G.J. (1963) *Crop Protection.* Leonard Hill, London.

Soper, F.L., Davis, W.A., Markham, F.S. and Riehl, L.A. (1947) Typhus fever in Italy, 1943–1945, and its control with louse powder. *The American Journal of Hygiene* 45, 305–334.

Taylor, W.A, Anderson, P.G. and Cooper, S. (1989) The use of air assistance in a field crop sprayer to reduce drift and modify droplet trajectories. *Proceedings of the Brighton Crop Protection Conference – Weeds*, 631–639. BCPC, Farnham, UK.

Tunstall, J.P., Matthews, G.A. and Rhodes, A.A.K. (1961) A modified knapsack sprayer for the application of insecticides to cotton. *Cotton Gowing Review* 38(1), 22–26.

Yeo, D. and Wilson, D.B. (1958) Aircraft application for anopheline control. *Annals of Tropical Medicine and Parasitology* 52, 402–416.

3 Insecticides Post-1950

Some of the insecticides developed in the mid-20th century have already been mentioned, but a number of important organophosphate and carbamate insecticides were developed and commercialized after 1960.

Organophosphates

Following the marketing of parathion and concerns about its toxicity, over the next six decades the agrochemical companies were looking for less toxic insecticides that could be readily absorbed by plants in which they could move systemically from the roots (soil/seed treatment) to protect young growth without a need to respray. Nevertheless, many were classified by the WHO as class Ia or Ib and regarded as highly hazardous.

The organophosphates used as insecticides can be divided into groups, including aliphatic (e.g. demeton-S-methyl, dimethoate), heterocyclic (e.g chlorpyrifos, triazophos) and phenyl (e.g. fenitrothion, parathion, temephos). These vary in their mammalian toxicity and persistence in the environment. They were used extensively following the banning of organochlorine insecticides.

Azinphos methyl

This extremely hazardous insecticide was first registered as Guthion in 1959 and became widely used in apple and other fruit orchards. Sprays were directed at codling moth larvae on apples. Despite its toxicity, sprays were applied with aircraft and air-assisted orchard sprayers in the apple-growing area in Washington State, where the effect of spray drift was subject to many studies. Ultimately, in the USA, the concern about

Table 3.1. WHO classification of pesticides.

WHO class		LD_{50} for the rat*	
		Oral	Dermal
Ia	Extremely hazardous	<5	<50
Ib	Highly hazardous	5–50	50–200
II	Moderately hazardous	50–200	200–2000
III	Slightly hazardous	>2000	>2000
U	Unlikely to present acute hazard	>5000	>5000

*mg/kg body weight

farm workers, pesticide applicators and aquatic ecosystems resulted in it being banned in September 2013 after a 12-year phase-out period. It had been banned in 2006 in the EU.

Chlorpyrifos

Chlorpyrifos was introduced by the Dow Chemical Company in 1965. It became very widely used as Dursban or Lorsban to control many different pests from locusts to domestic pests such as cockroaches, soil pests such as wireworms, and on cotton pests in Egypt and Pakistan. A study in China revealed that an inexperienced spray operator was likely to be exposed to eight times as much active ingredient as an experienced applicator spraying maize over 80 cm high. However, there were increasingly vocal concerns about its neurotoxic effect on humans, so in the USA, in June 2000, all homeowner uses were banned except when applied in baits to control ants and cockroaches. There were other restrictions on crops such as apples, citrus fruit and tree nuts, as well as tomatoes, and in 2012, application rates were lowered and buffer zones increased, especially around public places and residential areas. In the UK, all uses were revoked in March 2016, except for application to brassica seedlings via a gantry sprayer. This caused considerable concern with the operators of airports as chlorpyrifos was very effective against soil pests on the turf areas around runways, which required control to avoid birds being attracted to them and thus increasing the risk of bird strikes on aircraft.

Fenitrothion

Fenitrothion is a contact and stomach-acting insecticide that was introduced in 1960 by Sumitomo and became widely used on many crops and as a public health insecticide. Fenitrothion was used as a less toxic alternative to parathion, which was banned in some countries such as Japan. After the banning of dieldrin for locust control, it was included on a shortlist for control of desert locusts to provide a choice of OP with chlorpyrifos or malathion. It was also used extensively to control stored products pests and later mixed with a pyrethroid, possibly to provide protection over a longer period.

No longer recommended OPs

Several OP insecticides classified as WHO Ia or Ib were widely used in many countries but have now been withdrawn or are no longer manufactured. These include chlorfenvinphos, which was introduced in 1962 and, being a more persistent OP, was applied to control soil pests, although foliar sprays were used on potatoes to control the Colorado beetle. Its use was withdrawn in many countries including the USA, where its registration was cancelled in 1991. Similarly, methidathion, another non-systemic OP, introduced in 1965, competed with azinphos methyl on many crops, but relatively little was used in the USA. Methamidophos, introduced later in 1970, was widely used as both insecticide and acaricide on many different crops, presumably due to its systemic activity. Another systemic insecticide, monocrotophos, became widely used due to its relatively low cost. However, it was also highly toxic to birds and was implicated in human suicides, so its use became restricted. Its use on tomatoes and potatoes was withdrawn in 1985 and then all applications were discontinued in the USA in 1988. In 1952, diazinon, in WHO class II, was developed, and from 1955 was used extensively, for example in sheep dips to control blow flies, scab and lice disease. In 1988, in the UK, about 40 million sheep were dipped on over 18,000 farms. Unfortunately, many involved in sheep dipping became seriously ill, some with long-term neurotoxicity effects. Part of this was due to the design of the containers causing those preparing the dip to be exposed to the undiluted formulation that was invariably spilt on the top of the container. Those not wearing protective clothing were also exposed to the 'spray' created by the sheep emerging from the dip and vigorously shaking themselves to remove the liquid. Where sheep still need to be dipped, diazinon has largely been replaced by pyrethroids.

Less hazardous OP insecticides

Temephos, in WHO class III, became available in 1965 and has been extremely important in controlling larvae of vectors of human disease. Its most extensive use over a long period was its application to rivers in nine countries in the Sahel area of west Africa to control *Simulium* spp., the blackfly vector of onchocerciasis. Unfortunately, prolonged usage led to resistance being detected and it was replaced largely by the biopesticide *Bacillus thuringiensis israelensis* (Bti). Temephos is still being used in other river systems. Acephate is also in WHO class III and was introduced in 1972, principally to control aphids that were resistant to previously used insecticides. It has a moderate persistence of 10–15 days, and apart from application to vegetable crops was used on turf and to control ants. It was then replaced for aphid control by pirimiphos methyl, the carbamate

Table 3.2. Examples of organophosphate insecticides.

		LD_{50} for the rat*	
Insecticide	WHO class	Acute oral	Dermal
azinphos methyl	Ib	Approx. 9	150–200
chlorfenvinphos	Ib	10	31–108
chlorpyrifos	II	135 –163	>5000
demeton-S-methyl	Ib	Approx. 30	Approx. 30
dimethoate	II	387	>2000
fenitrothion	II	Approx. 1700	Approx. 810
methamdiphos	I	15.6	122
methidathion	I	25–54	297–1663
monocrotophos	Ib	18	130–250
naled	II	430	1100
parathion	Ia	Approx. 2	71–76
parathion-methyl	Ia	Approx. 3	Approx. 45
pirimiphos-methyl	III	1414	>2000
quinalphos	II	71	1750
temephos	III	4204	>4000
triazophos	Ib	57–59	>2000
trichlorfon	III	212	>5000

*mg/kg body weight

pirimicarb and, in some places, the neonicotinoid acetamiprid. Pirimiphos methyl, developed by ICI, has been used mainly in stored products pest control as it is relatively volatile and is applied as a fog. As mentioned earlier (see Chapter 2 – 'Formulation'), being volatile, it has also been reformulated as a microencapsulated suspension and used for indoor residual spraying against mosquitoes.

Carbamates

Aldicarb

In a glasshouse, with plants treated with potential herbicides, it was noted that all the flies had died, leading to the discovery that carbaryl was an insecticide; so Union Carbide examined other carbamates, and in 1965 announced the extremely toxic aldicarb. Perhaps the most toxic insecticide that was marketed, although only as a granule, it was used to treat a number of crops including potatoes, sugar beet and irrigated cotton, to protect against soil-borne pests, notably plant parasitic nematodes, but also as a systemic to control sucking pests. In the USA, when aldicarb was introduced, it was applied in the seed furrow at sowing to kill overwintered boll weevil and sucking pests for up to eight weeks after germination. It was considered for use on early-season trap crop in the boll weevil eradication programme, but cost and the perception of more bollworms on treated plants meant it was not used.

Ground maize cobs were initially used as the granule substrate in some trials, but further development sought a suitable formulation that was not attractive to birds. Much later, product stewardship and national regulations in some countries required the granule application to stop before the end of a row in case granules remained on the soil surface. In the USA, the EPA started limiting its use in 2010, aimed at complete withdrawal of the formulated granule by 2017. However, the same granule formulation has now entered US agriculture, initially in Georgia, as AgLogic 15G formulation, with strict stewardship. It had been argued that aldicarb had been used successfully for 40 years and its loss had led to economic losses on cotton and peanuts due to inadequate nematode control, for which aldicarb remains one of the most effective control agents.

Other highly hazardous carbamates

About the same time as aldicarb was marketed, Bayer introduced carbofuran, which is still used on many crops, including soybeans, for aphid control, but in the USA, since 2009, its use has been banned on any food crop. Similarly, in the EU it is now a banned insecticide. Later, a similar insecticide, methomyl, was introduced in 1967 by DuPont, and in 1974 oxamyl was marketed. Oxamyl is used as a 10% granule at the rate of 40–55 kg/ha in the furrow when planting potatoes. Lower rates are used with sugar beet, parsnips and carrots. In the UK, failure to return the container results in a charge of £25. All these competed with aldicarb to control nematodes, so were primarily available as a granule formulation to be used at planting. A key advantage of aldicarb and oxamyl as nematicides is their lower lipophilicity compared with OPs, resulting in greater mobility in soils without the need for rotavation following application. The use of all these has raised questions about their persistence and movement in soils to groundwater. A survey in the USA showed that oxamyl was detected in less than 0.1% of approximately 13,000 water systems examined, which led to its withdrawal in a number of states. Methiocarb has been used as a molluscicide since 1962, as pellets or granules.

Less hazardous carbamates

Pirimicarb was developed in 1969 as a very selective aphicide for use on cereals, vegetables and in orchards. In 1977, Union Carbide introduced thiodicarb as a stomach poison, although when applied as a seed treatment it is translocated systemically through the plant. It was also included in the cotton recommendations in Zimbabwe referred to earlier (see Chapter 1). This was in part a replacement for DDT but also as an alternative to pyrethroids. Thiodicarb has also been used as a molluscicide.

In 1979, carbosulfan was developed by FMC and is in WHO class III. When asked in an African country why they had not registered carbosulfan, the reply was that the EPA had not registered it. No doubt in the USA the company was not too concerned as the more hazardous carbamate was still being used.

The next stage has been to develop insecticides with different modes of action, rather than analogues of existing insecticides.

Table 3.3. Examples of carbamate insecticides.

		LD_{50} for the rat*	
Insecticide	WHO class	Acute oral	Dermal
aldicarb	Ia	0.93	20
bendiocarb	II	40–150	566–800
carbaryl	III	500–800	>4000
carbofuran	II	Approx. 8	>2000
carbosulfan	II	185–250	>2000
methiocarb	II	Approx. 20	>5000
methomyl	Ib	34	>2000
oxamyl	Ib	3.1	>2000
pirimicarb	II	142	>2000
thiodicarb	II	66	>2000

*mg/kg body weight

Pyrethroids

The story about pyrethroid insecticides began in the 1930s when scientists were frustrated by the lack of persistence of the natural pyrethrins, especially once a deposit was exposed to sunlight. It was not until 1949 that a synthetic version was made, namely allethrin I and II, with similar properties to the natural pyrethrins, especially the quick knock-down effect on insects. However, Charles Potter, who had moved to Rothamsted, decided that his department should get involved in altering the molecular structure of pyrethrin, so subsequently, Michael Elliott synthesized resmethrin in 1962, followed by bioresmethrin in 1967, although neither of these were stable in sunlight. There was much criticism of a government funded research station doing what it was considered the commercial industry should have been doing. Potter persisted, and with the government setting up the National Research Development Corporation to commercialize the outputs of this research, success came with Elliott developing permethrin in 1972, which was the first synthetic pyrethroid that was photostable (Elliott, 1976). Potter would have liked this to be retained for public health use but it was licensed to the agrochemical industry. At the same time the Japanese company Sumitomo began doing similar studies, and in 1978, fenvalerate was discovered. The (S) enantiomer of fenvalerate – esfenvalerate – is also marketed. Meanwhile, Rothamsted went on to develop more powerful pyrethroids, namely cypermethrin and deltamethrin, both becoming major insecticides in the global market. At the low dose per hectare they are relatively less hazardous to use than organophosphates and less persistent in the environment than the chlorinated hydrocarbons. Industry subsequently, in 1984, developed beta-cyfluthrin, lambda cyhalothrin, etofenprox and bifenthrin (Schleier and Peterson, 2011).

The use of organochlorine insecticides on cereals in the UK had declined by 1982 and was replaced by organophosphates, which were then replaced by pyrethroids, requiring a much smaller quantity to be applied per hectare (Fig. 3.1), helping to continue increasing cereal yields (Fig. 3.2).

Table 3.4. Examples of pyrethroid insecticides.

		LD_{50} for the rat*	
Insecticide	WHO class	Acute oral	Dermal
beta cyfluthrin	Ib	>77	>5000
bifenthrin	II	54.5	>2000
bioresmethrin	III	7000–8000	>10,000
permethrin	II	430–4000	>2000
cypermethrin	II	250–4150	>4920
deltamethrin	II	135–>5000	>2000
etofenprox	III	>42,000	>2000
fenvalerate	II	451	1000–3200
lambda-cyhalothrin	III	>11,000	>5000
resmethrin	III	>2500	>3000

*mg/kg body weight

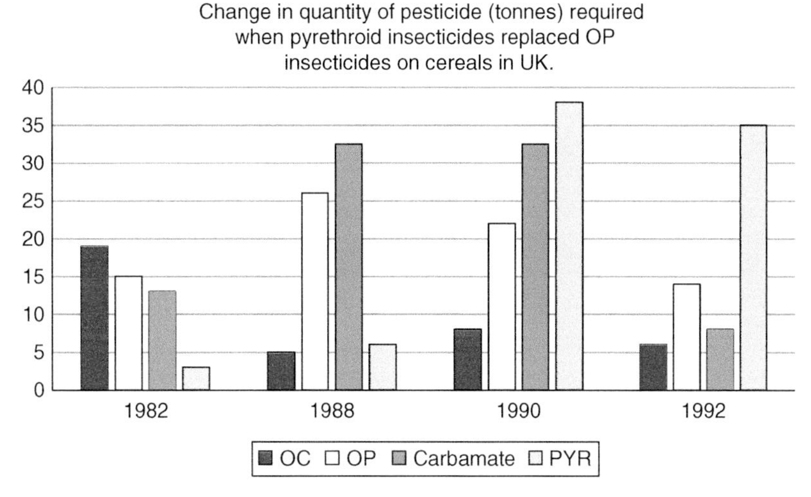

Fig. 3.1. Change in use of insecticides on cereals. (Redrawn from Wilson (1995), reproduced by permission of the Royal Society of Chemistry.)

In Africa, pyrethroids, notably fenvalerate, lambda cyhalothrin and deltamethrin, replaced DDT in the spraying programme against bollworms, but in Zimbabwe they could only be applied when scouting indicated a need for control of *Helicoverpa armigera*.

These developments came at a most appropriate time, as concerns raised in 1962 by Rachel Carson in her book *Silent Spring* were beginning to have an effect with the banning of DDT. Although the photostability of pyrethroids allowed effective use on crops, they did not accumulate in the food chain. This was good for the agrochemical industry, but it was unfortunate that they became too extensively used, with 33 million ha treated annually and occupying 25% of the global insecticide market, which inevitably led to insects that were resistant to pyrethroids being detected. Large boom sprayers are now being used in the UK. (Figs 3.3 and 3.4).

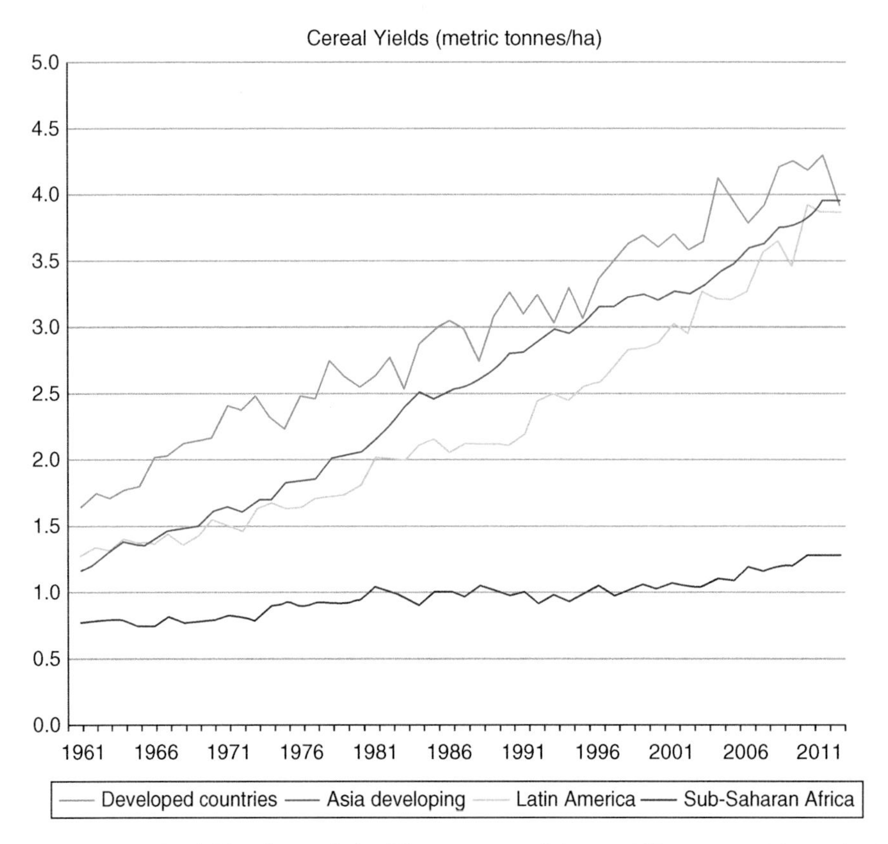

Fig. 3.2. Increase in yields of cereals in different parts of the world between 1961 and 2011. (Redrawn from World Resources Institute.)

Fig. 3.3. Househam tractor trailed sprayer in UK. (Photo courtesy of Househam.)

Fig. 3.4. Househam self-propelled sprayer. (Photo courtesy of Househam.)

Meanwhile, Chris Curtis decided to investigate whether treating bed nets with pyrethroids would be a more effective way of protecting people from malaria. Initially, the individual bed nets were treated by soaking a net in the diluted insecticide and dried, which could be done by villagers. Concerns about this process led to impregnating the pyrethroid in the fibres used to make the bed nets. Insecticide treated nets (ITN) became long-lasting insecticide treated nets (LLIN). Permethrin, considered by Potter to be ideal for use in public health, was the first to be used, but later, other pyrethroids were also impregnated into nets. An estimated 655,000 deaths occurred as a result of malaria infection in 2010 and about 86% of the cases occurred in children under 5, with some 91% of malaria deaths occurring in the WHO Africa region. In Africa, the proportion of households with one or more ITNs or LLINS had increased to 79% by 2015, with a highly significant impact on malaria. Mortality in children under 5 and protected by nets was reduced by about 50%. A small study in Cameroon revealed that adults continued to have malaria as they do not go to bed so early and go outside houses during the evening or early morning when the mosquito vector is still active.

The main problem with the treated bed nets was that when a high proportion of houses in a village are using the nets, the mosquito population is continually exposed to the insecticide, throughout the year. This has inevitably led to selection of vectors resistant to pyrethroids. Pyrethroid in the net can be effective for up to five years, although many nets will be torn and have holes that allow mosquitoes to enter after about three years. The difficulty is finding an alternative suitable low-toxicity insecticide that can be impregnated on a net. Currently, some manufacturers are adding piperonyl butoxide to the net so that pyrethroid-resistant mosquitoes are also killed. A combination of alpha cypermethrin and chlofenapyr on a polyester net has now been given interim approval.

Recalling the effectiveness of indoor residual spraying with DDT, there is also greater awareness of spraying walls of houses, as there are insecticides with different modes of action. So far, relatively little attention has been given to other methods of vector control in Africa, where housing standards are poorer than in many countries and where screening of houses and control of outdoor vector populations has also had a major effect, enabling malaria to be eliminated. With more people active outdoors after sunset, greater attention is needed to control the mosquito populations outside houses.

The establishment of the Innovative Vector Control Consortium (IVCC) based at the Liverpool School of Tropical Medicine was in recognition that the production of insecticides for vector control was such a small fraction of the agrochemical market that another model was needed to examine and bring to market new insecticides with different modes of action. Initially, this approach has enabled some new formulations of old chemistry to be developed. One example is pirimiphos methyl, evaluated by the WHO in the 1970s but considered too volatile for indoor residual spraying, which has now been used as a micro-encapsulated formulation that prolongs its activity for several months. Another approach has been to evaluate a neonicotinoid insecticide, clothianidin, for indoor residual spraying.

Tefluthrin was introduced by ICI in 1986 for application to soil to control various pests in the root zone of maize, sugar beet and other crops, because it is more volatile than other pyrethroids and thus is distributed in soil air spaces. Tetramethrin is another contact pyrethroid with rapid knock-down, introduced in Japan in 1964, which has been used as a public health insecticide and in garden use.

Neonicotinoids

Neonicotinoid insecticides have been developed since the late 1980s, principally as systemic insecticides with a similar mode of action to nicotine. They selectively bind and interact with the insect nicotinic acetylcholine receptor site and cause paralysis, which leads to death, often within a few hours, but are much less toxic to mammals. Rapidly, they became the main commercial insecticide where insect pests were resistant to the pyrethroids. Being systemic, a key use of these insecticides was a seed treatment for many crops including cotton, on which imidacloprid was used to control sucking pests prior to when the plants started bud protection, and required monitoring to determine if a spray against bollworms was required. The rise in their use in the USA is documented by Douglas and Tooker (2015).

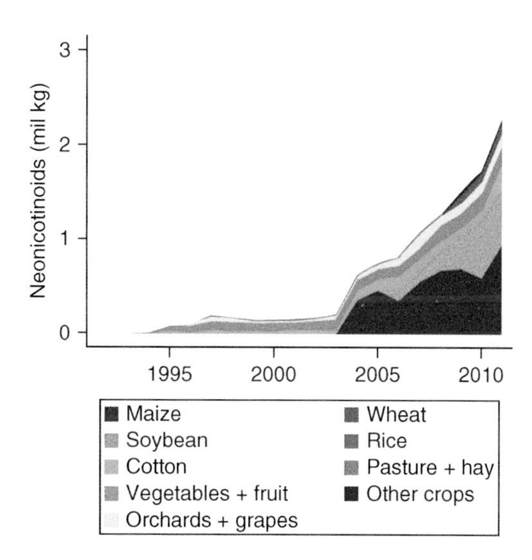

Fig. 3.5. The quantity of neonicotinoids used globally on different crops. (From Douglas and Tooker, 2015.)

However, poor seed treatment in 2008 led to the death of a vast number of bees being reported in Germany's Baden-Württemberg state. This resulted in the immediate suspension of the approval for eight seed-treatment products used in oilseed rape (canola) and sweetcorn, which contained imidacloprid, clothianidin and thiamethoxam. Most bees were apparently killed by clothianidin due to a poor formulation that meant the insecticide had not been glued sufficiently to the seeds. The problem was made worse by the use of seeders that vented dust from the seeds up into the atmosphere, so the insecticide dust drifted downwind. Improved formulation of the insecticide on seeds and modification of the seeders prevented dust being released in the environment. Nevertheless, the EU imposed a moratorium on the use of neonicotinoids on the basis that their use was killing bees. Some countries requested a derogation so that they could continue to use treated seed, as the alternative was spraying other insecticides, which could be more detrimental to bees, or to stop growing oilseed rape, which can suffer severe crop damage due to cabbage stem flea beetles during early crop growth. The outcome of the moratorium within the UK has cost the farmers growing oilseed rape an estimated £18.4 million and resulted in the loss of 28,800 ha of crops, even though farmers increasingly turned to alternative pesticides to control cabbage stem flea beetle (CSFB) (Scott and Bilsborrow, 2017).

In the USA, in 2012, the EPA denied a petition to suspend the use of clothianidin made by several anti-pesticide organizations. The following year the same group and others again petitioned the EPA, accusing the agency of performing inadequate toxicity evaluations and allowing insecticide registration based on inadequate studies. Then, later the same year, a Save American Pollinators Act was introduced in the House of Representatives, calling for suspension of the use of four neonicotinoids, including the three recently suspended by the EU, until their review was complete. It was then assigned to a congressional committee.

Bees in some parts of the world have undoubtedly suffered from what is referred to as 'colony collapse disorder'. The exact cause of the decline in bees is not clear, but it did not occur in Australia where neonicotinoids continued to be used and where the varroa mite is not present. Where varroa mite does occur, it is one of the causes of population decline, but other factors include viruses transmitted by the mite, a lack of wild flowers, detrimental weather conditions and, in some cases, poor husbandry of bee colonies, especially when hives are moved to provide pollination on different farms. Much of the argument against using neonicotinoids is based on laboratory or other small-scale investigations, where the dose of insecticide available to test beehives is higher than would be found on treated crops. There is a clear distinction between small-scale experiments in the laboratory and larger field-scale trials in terms of effect on bees and the dose applied. Thus, at the doses used in many experiments where there is a definite effect, it is not observed in the field at the range of maximum residues of neonicotinoid insecticides detected in nectar on spring- and autumn-sown oilseed rape crops (Fig. 3.6). Environmentalists have also considered that the neonicotinoids have had an adverse impact on bumblebees and other wild bee populations.

As with other main groups of insecticides, there are considerable variations between individual neonicotinoids. The group can be divided into

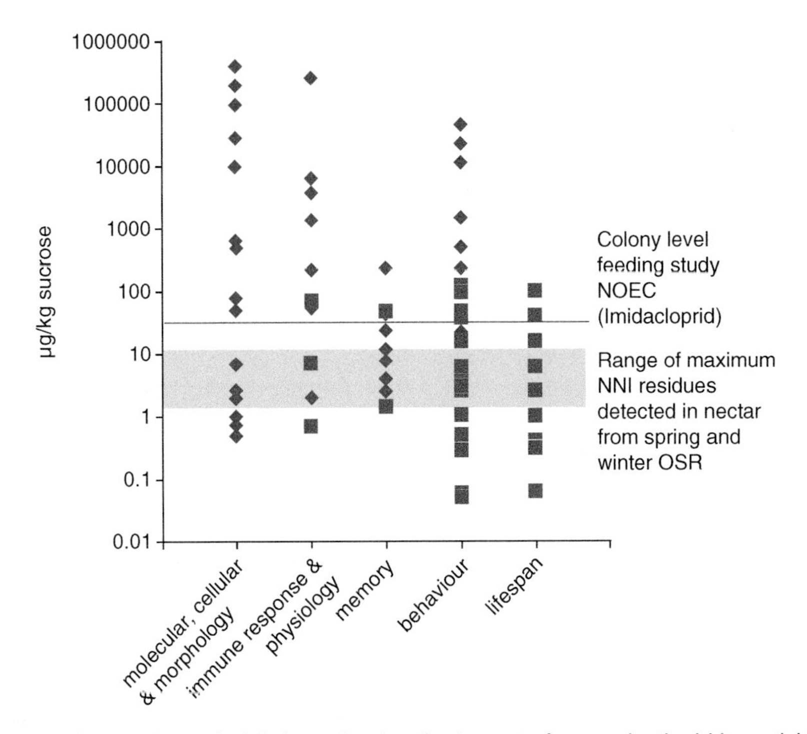

Fig. 3.6. Comparison of trials investigating the impact of a neonicotinoid insecticide (imidacloprid) on bees. (Redrawn from Walters, 2016.)

Table 3.5. Examples of neonicotinoid insecticides.

		LD_{50} for the rat*	
Insecticide	WHO class	Acute oral	Dermal
acetamiprid	II	146–217	>2000
imidacloprid	II	Approx. 450	>5000
clothianidin	III	>500	>2000
dinotefuran	III	2000–2800	
sulfoxaflor	III	1000	
nitenpyram	III	1680	>2000
thiacloprid	II	396–836	2000
thiamethoxam	III	1563	>2000

*mg/kg body weight

four types based on their chemistry: chloropyridyl (e.g. imidacloprid), thiazoyl (e.g. thiamethoxam), furanyl (e.g. dinotefuran) and sulfoximine (e.g. sulfoxaflor). Triacloprid is the first chloronicotinyl insecticide to show activity against weevils, leafminers and various species of beetles, as well as against sucking insects such as aphids, whiteflies and some jassids. It has also shown good control of pests within the upper plant foliage after soil application.

Neonicotinoids were developed due to their low mammalian toxicity and good systemic activity, which, theoretically, makes them far superior to many of the insecticides previously developed. However, the incident reported above demonstrates that the whole process of delivery of an insecticide in the field has to be designed so that adverse effects on non-target species are avoided (Walters, 2016). Recently, it has been pointed out that the sensitivity of honey bees to neonicotinoids varies by orders of magnitude, with bees being less sensitive to N-cyanoamidine compounds, such as thiacloprid (Manjon *et al.*, 2018).

Phenyl-pyrazole Insecticides

Fipronil, discovered in 1987, was developed originally by Rhône-Poulenc and initially promoted as an alternative to control the desert locust. Trials were carried out in west Africa and it was shown to be very effective against acridids. FAO discontinued recommending blanket treatments, even at a dose of 4 g a.i/ha, as its use in Madagascar over an extensive area had a devastating impact on the harvester termites (*Coarctotermes clepsydra*) and other non-target organisms (Peveling *et al.*, 2003). However, its use continued in Australia where ULV spraying at very low dose rates was used to control the mobile bands of plague locust *Chortoicetes terminifera* hoppers. Barrier treatments applied swaths at 300-m intervals for an overall dose of 0.33 g a.i./ha. Studies indicated no impact on common wood-eating *Microcerotermes* spp. (Maute *et al.*, 2016).

Post-1990 Insecticides

Some of the following newer insecticides, in some cases microbe- or plant-derived, were developed specifically to control sucking pests such as aphids and whiteflies, rather than lepidopteran pests. This may have been considered necessary as these pests are often on the underside of leaves, so less exposed to direct effects of sprays when applied either by booms on tractors or aircraft. They tend to be more mobile in plants. Their use is also very important where crops containing the genetic modification to express Bt toxins are grown.

Oxadiazine

Indoxacarb, introduced around 2000, is a WHO class III insecticide. It acts by blocking the nerve sodium channels, but at a different binding site to that with pyrethroid insecticides, so cross-resistance was not expected. It has been marketed to control lepidopteran pests but it is active against a wide range of insect pests, although slower acting on some sucking pests (King *et al.*, 2000). It is also sold in products to control household pests such as cockroaches and ants and it is also effective against fleas on pets.

Pyridine

Flonicamid

This has been developed since the late 1990s as a low toxicity systemic aphicide that fits well in IPM as beneficial insects are not affected. It inhibits feeding by aphids within 0.5 hours of treatment, resulting in death by starvation. It is also active against other sucking pests such as whiteflies (Morita *et al.*, 2007).

Pymetrozine

This is another low toxicity aphicide developed since 1993, which is used on vegetable crops and ornamentals, and also to control whiteflies and pollen beetles.

Pyrroles

Chlorfenapyr, introduced in 2001, is a pro-insecticide (i.e. it is metabolized into an active insecticide after entering the insect pest). It was derived from a class of compounds produced from microbes, known as halogenated pyrroles. The active ingredient is a WHO class III insecticide. It is slow-acting but is now used on non-food crops. Chlorfenapyr is used as a wool insect-proofing agent and is also being evaluated for use in mosquito control (Raghavendra *et al.*, 2011).

Pyrazole

Tebufenpyrad is similar to rotenone and is a strong mitochondrial complex inhibitor. Since 2002 it has been used mainly in greenhouses around the world and is registered in the USA for use on ornamental plants in commercial greenhouses.

Abermectin

This acaricide/insecticide, introduced in 1985, was derived as a fermentation product from *Streptomyces avermitilis*, following the discovery of the anthelmintic ivermectin in the late 1970s, which led to the award of a Nobel prize to William Campbell in 2015 for his contribution to the control of 'river blindness'. Emamectin benzoate, based on a naturally occurring soil actinomycete, has also been effective in controlling many insect pests. Other acaricides are discussed later in this chapter.

Spinosyns

A new class of insecticide, the naturalyte class was developed from fermentation products obtained from *Saccharopolyspora spinosa* found in a soil sample from the Caribbean in 1982. There are over 20 natural forms of spinosyns, and over 200 spinosoids have been synthesized. Spinosyns and spinosoids have a novel mode of action, primarily targeting binding sites on nicotinic acetylcholine receptors (nAChRs) of the insect nervous system that are distinct from those at which other insecticides have their activity.

Spinosad contains a mix of two spinosoids, spinosyn A (the major component) and spinosyn D (the minor component), in a roughly 17:3 ratio. It was first registered in the USA in 1997 and is used globally to control a wide range of pests. In the UK, it is used mostly on vegetable crops with a maximum of four treatments per crop on brassicas, onions and leeks, and one pre-blossom and three post-blossom on apples and pears. There is an extension of authorization for its use on minor crops (EAMU).

Anthranilamides/diamides

These insecticides are refinements of the botanical insecticide ryania and function by activating the ryanodine receptors in intracellular calcium channels in larvae and adult insects that ingest them. This causes a large release of calcium ions and, consequently, muscle contraction so that the insect dies (Satelle *et al.*, 2008).

Chlorantraniliprole, introduced in 2007, has low mammalian toxicity and is especially effective against lepidopteran pests, but also against beetles, including the Colorado beetle (Lahm *et al.*, 2007). It is registered in the UK for control of codling moth on apples and pears and is 'off-label' on some other crops, such as hops, as part of an IPM programme.

Cyantraniliprole is recommended in the UK, except when crops are flowering, as part of an IPM programme, e.g. on cabbage root fly on broccoli, brussels sprouts, cabbages and cauliflowers. It is highly toxic to bees.

Flubendiamide, originally reported in 1979, was registered in the USA for use on over 200 crops with some crops having as many as six applications per year, but in 2016 the EPA concluded that the continued use of flubendiamide would result in unreasonable adverse effects on the environment, particularly benthic invertebrates, which are an important part of the aquatic food chain, particularly for fish.

Pyridalyl

Pyridalyl, which was first used in Japan in 2004, has a low mammalian toxicity. It has a unique chemical structure unrelated to other insecticide groups. It is very effective against lepidopteran pests, with anti-feedant effects, rapidly resulting in less damage. It is also effective against thrips, including western flower thrips (*Frankliniella occidentalis*), but harmless to natural enemies and pollinators, so is expected to be very important in IPM programmes (Nishimura *et al.*, 2007).

Ketoneols

This new group of insecticides is based on the interruption of the biosynthesis of lipids in insects.

Spiromesifen, a new WHO class III insecticide, developed in 2002, is in a group of spirocyclic phenyl-substituted tetronic acids. It is effective as a foliar contact and translaminar insecticide against whiteflies and spider mites. It acts as a lipid biosynthesis inhibitor that affects the immature stages more rapidly than adults. Spray deposits can persist on foliage for 14–21 days, with a pre-harvest interval of three days. Spiromesifen was considered to be suitable in IPM programmes as laboratory and field tests showed it to be safe on beneficial insects.

Spirotetramat is active against aphids, mites and whiteflies. Apart from its effect on immature stages, it also reduces the fecundity of adult female sucking pests. It is a systemic insecticide that is sprayed on the leaves of the plant, penetrates them and moves downwards so that it also protects the roots (Nauen *et al.*, 2008). In the UK it is used off-label on fruit and vegetable crops.

Spirodiclofen is similar, with activity against mites, psyllids and scale insects. In the USA it has been applied on citrus, vines, nut trees and other fruit trees. Similarly, in the UK it is used mostly on fruit crops and some vegetables, e.g cucumbers. Multi-resistant strains of spider mites were susceptible to spirodiclofen, but no doubt care will be needed in how extensively it is used (see 'Resistance' in Chapter 7).

The activity of the ketoenols complements the use of Bt toxin in GM crops by controlling insect pests unaffected by Bt.

Insect Growth Regulators

There are hormonal IGRs that mimic or inhibit the juvenile hormone that affects the moulting of the insect as it progresses from a young larva towards pupation, for example causing premature moulting, or preventing a pupa becoming an adult. Other IGRs act by inhibiting another hormone, ecdysone, thus causing the larva to moult into a larger larva and not into a pupa. They have been effective against the nymphal stages of sucking pests such as whiteflies. Other compounds, the benzoylureas, impair the formation of new cuticles at moulting. IGRs are of low mammalian toxicity and are generally slow-acting, which makes some users apprehensive as to whether they will be able to control a pest.

Methoprene, developed from 1973 as a juvenile hormone insecticide, has been used for control of mosquito larvae. Pyriproxifen, a juvenile hormone analogue, is a pyridine insecticide, developed in 1989, that has been very successful for controlling mosquito larvae at very low rates of application. Studies have shown that adult mosquitoes inside houses can pick up this insecticide and redistribute it while ovipositing to control larvae.

Diflubenzuron, a chitin deposition inhibitor, which was developed in 1972, has been applied in many different crops and in forestry against mostly lepidopteran larvae. It has also been recommended to control desert locust hoppers and immature stages of mosquitoes. Although it does not kill adult insects, there is evidence that oviposition is reduced. Other benzoylurea insecticides are chlorfluazuron, flufenoxuron (now banned within the EU due to potential of bioaccumulation in the food chain), hexaflumuron and triflumuron. Lufenuron was developed and is used mainly against fleas on pets.

Teflubenzuron, developed in 1983, is an acyl-urea insecticide that has also been used to control lice on fish. Tebufenozide, developed in 1992, is unusual as it acts as an anti-juvenile hormone, causing larvae to try to moult into precocious adults. It has been used against forest pests.

Biopesticides

Development of biopesticides has been traditionally confined to small companies, but with integrated pest management now part of the EU's policy on sustainable pesticide use, some of these companies are now a part of the agrochemical/plant science industry. Thus there is a mechanism for existing products to be marketed on a much larger scale, and further research should enable other potential biopesticides to proceed from laboratory tests to field-scale implementation. The development of the mycoinsecticide to control locusts illustrated the gap between a potential insecticide and one that could be formulated and applied on a large scale. Biopesticides are also discussed in Chapter 8.

Dieldrin had been the insecticide selected for control of the desert locust, but following the drive to ban persistent organic pesticides (POPs), a research project examined the potential of using a mycoinsecticide based on a *Metarhizium* to control locusts. A sample of *Metarhizium* derived from a locust was in a collection at Kew (IMI330189) so CABI set up a research programme (known as LUBILOSA), which operated for ten years. Apart from a search for other fungi on locusts, a method of culturing the existing sample, later recognized as *M. acridum,* and separating the spores from the substrate was developed. The next stage was translating a source of fungal spores so that these could be applied using ultra-low-volume spraying techniques needed for the control of desert locusts in arid areas of Africa. The lipophilic spores could be suspended in oil, and subsequent field trials showed that ULV sprays, initially with a low-tech formulation using a mixture of vegetable oil (30% peanut oil) and kerosene as a carrier of the spores, or a mixture of mineral oils, were feasible (Bateman *et al.*, 1998; Lomer *et al.*, 2001). Examining the data from a large-scale aerial spraying trial, with plots sprayed with the mycoinsecticide Green Muscle, an OP insecticide (fenitrothion), and an untreated area, it was found that the initial rapid decline in locusts following the OP spray was followed by immigration of more locusts, whereas birds were unaffected by the mycoinsecticide and ate the moribund locusts; so the population declined more slowly, with 70–90% dead within 14–20 days. Although more expensive than an OP, there was no environmental cost of using the mycoinsecticide, nor was there a cost of processing obsolete stocks as the spores degraded naturally when removed from temperature-controlled stores. Later, FAO introduced a priority for choosing an insecticide for locusts, so where possible, the mycoinsecticide Green Muscle was used unless urgent rapid action was essential. Similarly, in Australia, the locally developed Green Guard has been used against the plague locust. *Metarhizium anisopliae* is now used in the UK off-label to control vine weevils on strawberry and other soft-fruit crops.

The use of *Bacillus thuringiensis* had continued since the 1950s. Some of the new formulations contain potent strains of *Bacillus thuringiensis* subspecies *aizawai* (Bta) to control caterpillar pests on vegetables, fruits, nuts, row crops and turf. One product features a balanced blend of four potent toxin proteins and a spore, targeting key pests such as armyworms and diamondback moth larvae. *Bacillus thuringiensis kurstaki* is also widely used in forests and vegetable crops. Timing of sprays and application were important as the biopesticide has to be ingested, ideally by the younger instars before damage is done. The advantage of the genetically modified crops incorporating the Bt toxin is that it is ingested by the first instar larvae as they start to bite plants.

Beauvaria bassiana is marketed as a biopesticide for use in glasshouses to control glasshouse whitefly (*Trialeurodes vaporariorum*) on vegetables and ornamentals. Spores are provided as a 7% suspension in oil and applied at 3 l/ha. It is an ideal complement to integrated pest management programmes with minimal use restrictions and a favourable toxicological and environmental profile to give flexibility to the grower.

Baculoviruses

A number of the viruses isolated from insect pests have been considered as possible control agents. One problem has been the need to culture the virus in living insects, but some have been marketed as commercial products. Commercially available baculoviruses include the codling moth granulosis virus, gypsy moth (*Lymantria dispar*), the leafworm (*Spodoptera* spp.) and cassava hornworm (*Erinnyis ello*) granulosis virus (Lacey *et al.*, 2015).

Botanical Insecticides

Although the agrochemical industry has developed insecticides based on chemicals extracted from plants with long-established pesticidal activity, there has been an interest in still using these botanical extracts and seeking other plants having similar activity. The main problem is the variability in the insecticidal chemicals within different parts of the plants or between species.

Belmain *et al.* (2012) have reported that not all rotenoids in *Tephrosia vogeli* are equally effective and that the occurrence of rotenoids in leaves shows substantial variation during the growing season. Stevenson *et al.* (2017) have suggested that there is increasing interest in botanical pesticides with the adoption of IPM in crop protection policy. Whether the indigenous knowledge of pesticidal plant resources among poor small-scale farmers who were involved with the production of pyrethrum in east Africa can be developed depends on whether resources are made available to overcome the many hurdles in developing a new generation of cash crops for botanical pest management.

Table 3.6. Key plants from which insecticides have been extracted.

Pesticidal plant	Pesticide	Target pests
Chrysanthemum cineraria	Pyrethrum	Mosquitoes and other insects
Tagetes minuta	Mexican marigold	Aphids, etc. and nematodes
Tephrosia vogeli	Rotenone	Pests of stored grain
Azidirachta indica	Neem	Various

Entomopathogenic Nematodes (EPNs)

EPNs belong to the closely related genera *Steinernema* and *Heterorhabditis* and have an important role, specifically to control soil pests where moisture is sufficient for the nematodes to survive and seek out the pests. Nematode infective juveniles (IJ) swim freely through moist compost in water films. They are attracted to insect larvae and enter the body where mutualistic bacteria in the nematode gut are released and kill the pest organism within 24–48 h. *Steinernema kraussai* is one example that can be cultured on a large scale and can enable the infective juveniles to be released by spraying or via an irrigation system. The nematodes enter the vine weevil (*Orthorhinus klugi*) larvae, release bacteria and feed and reproduce on the resulting 'bacterial soup' in the insect cadaver, releasing new IJ to attack other larvae once resources in the insect cadaver are exhausted. Similarly, *Steinernema feltiae* has been used to control sciarid flies (*Bradysia* spp.) and other insects, including western flower thrips, leafminer and scale insects on young seedlings in glasshouse production, such as herbs (Wright *et al.*, 2005).

On a small scale, the EPN can be mixed with water and poured as a spot treatment near plants liable to be attacked. Spraying is also possible, but care is needed where plants are in individual pots in trays to ensure even distribution of the EPNs to all pots. The dose is 0.5–1 million IJs/m^2.

A related nematode species, *Phasmarhabditis hermaphrodita*, is marketed as Nemaslug to control slugs on horticultural crops.

Biocontrol of Plant Nematodes

Where cyst nematodes are a pest of vegetable and ornamental crops, there is now interest in biological control using *Pasteuria* spp. For example, *Pasteuria nishizawae* is a mycelial and endospore-forming bacterium parasitic on cyst nematodes of genera *Heterodera* and *Globodera*. *Pasteuria nishizawae* is now marketed by Pasteuria Bioscience, now part of Syngenta, to control the soybean cyst nematode. Root-knot nematodes, *Meloidogyne* spp., are important parasitic nematodes of vegetable and ornamental crops. Positive results have been achieved with *Pasteuria penetrans* for the control of *Meloidogyne incognita* by seed treatment of cucumber (Kokalis-Burelle, 2015).

Another biopesticide, marketed to control plant parasitic nematodes, is based on the beneficial fungus *Paecilomyces lilacinus* (synonym *Purpureocilium lilacinum* strain 251). Eggs and larvae of *Meloidogyne* spp., burrowing nematodes (*Radopholus similis*), *Globodera* spp. and root lesion nematodes (*Pratylenchus* spp.) are affected before they can attack the roots. Similarly, a combination of *Bacillus subtilis* and *Bacillus licheniformis* is now recommended for the control of *Pratylenchus zeae* and *Meloidogyne incognita* on several crops in Brazil.

Acaricides

Older acaricides, including amitraz and tetradifon, are noted in Chapter 1. Azocyclotin, an organotin, introduced in 1977, is now obsolete and no longer manufactured. Apart from spirodiclofen and spiromesifen, mentioned above, several new acaricides, including acequinocyl, bifenazate, cyflumetofen, etoxazole, fenazaquin, fenpyroximate and hexythiazox have been developed. Contact and residual effects of these on natural enemies has been conducted to determine their role in IPM. Where they have contact activity, distribution of the spray to where mites are located within the crop canopy is critical.

Hexythiazox, a carboxamide, was first registered in 1985 and is used in the EU but not the UK, and used to control tetranychid mites. Acequinocyl was first noted in 1990 in Asia and has been registered in some EU countries. Fenpyroximate, a pyrazolium, was introduced in 1991. It is registered within the EU, excluding the UK. Fenazaquin, a quinazoline, was introduced in 1993 and may be used in the UK, although it is not in the main listing of products. The hydrazine carboxylate bifenazate was introduced in 1999 and is registered within the EU. In the UK it has been registered for use on strawberries with a maximum of two applications per year. Similarly, etoxazole, a diphenyl oxazoline contact acaricide, was first registered in 1998. Cyflumetofen, a bridged diphenyl, was introduced in 2004 and has been used in Holland and Belgium.

References

Bateman, R.P., Douro-Kpindu, O.K., Kooyman, C., Lomer, C. and Oambama, Z. (1998) Some observations on the dose transfer of mycoinsecticide sprays to desert locusts. *Crop Protection* 17, 151–158.

Belmain, S.R., Amoah, B.A., Nyirenda, S.P., Kamanula, J.F. and Stevenson, P.C. (2012) Highly variable insect control efficacy of Tephrosia vogelii chemotypes. *Journal of Agriculture Food Chemistry* 60, 10055–10063.

Carson, R. (1962) *Silent Spring*. Houghton Mifflin Co., Cambridge, Massachusetts.

Douglas, M.R. and Tooker, J.F. (2015) Large-scale deployment of seed treatments has driven rapid increase in use of neonicotinoid insecticides and preemptive pest management in U.S. field crops. *Environmental Science & Technology* 49, 5088–5097.

Elliott, M. (1976) Properties and applications of pyrethroids. *Environmental Health Perspectives* 14, 3–13.

King, K.D., Sader, M., Kagaya, Y., Tsurubuchi, Y., Mulderig, L., Connair, M. and Schnee, M. (2000) Bioactivation and mode of action of the oxadiamine indoxacarb in insects. *Crop Protection* 19, 537–545.

Kokalis-Burelle, N. (2015) *Pasteuria penetrans* for control of *Meloidogyne incognita* on tomato and cucumber, and *M. arenaria* on snapdragon. *Journal of Nematology* 47, 207–213.

Lacey, L.A., Grzywacz, D., Shapiro-Ilan, D.I., Frutos, R., Brownbridge, M. and Goettel, M.S. (2015) Insect pathogens as biological control agents: back to the future. *Journal of Invertebrate Pathology* 132, 1–41.

Lahm, G.P., Stevenson, T.M., Selby, T.P., Freudenberger, J.H., Cordova, D. *et al.* (2007) Rynaxypyr: a new insecticidal anthranilic diamide that acts as a potent and selective ryanodine receptor activator. *Bioorganic & Medicinal Chemistry Letters* 17, 6274–6279.

Lomer, C.J., Bateman, R.P., Johnson, D.L., Langewald, J. and Thomas, M. (2001) Biological control of locusts and grasshoppers. *The Annual Review of Entomology* 46, 667–702.

Manjon, C., Troczka, B.J., Zaworra, M., Beadle, K., Randall, E. *et al.* (2018) Unravelling the molecular determinants of bee sensitivity to neonicotinoid insecticides. *Current Biology* 28, 6. Available at: https://doi.org/10.1016/j.cub.2018.02.045 (accessed 28 March 2018).

Maute, K., French, K., Story, P., Bull, C.M. and Hose, G.C. (2016) Effects of two locust control methods on wood-eating termites in arid Australia. *Journal of Insect Conservation* 20, 107–118.

Morita, M., Ueda, T., Yoneda, T., Koyanaqi, T. and Haqa, T. (2007) Flonicamid, a novel insecticide with a rapid inhibitory effect on aphid. *Pest Management Science* 63, 960–973.

Nauen, R., Reckmann, U., Thomzik, J. and Thielert, W. (2008) Biological profile of spirotetramat (Movento) – a new two-way systemic (Ambimobile) insecticide against sucking pest species. *Bayer Crop Science Journal* 61, 2.

Nishimura, S., Saito, S. and Isayama, S. (2007) Pyridalyl: a novel compound with excellent insecticidal activity, high selectivity and unique mode of action. *Japanese Journal of Plant Science* 1 (2), 85–94.

Peveling, R., McWilliam, A.N., Nagel, P., Rasolomanana, H., Raholijaona *et al.* (2003) Impact of locust control on harvester termites and endemic vertebrate predators in Madagascar. *Journal of Applied Ecology* 40, 729–741.

Raghavendra, K., Barik, T.K., Poonam Sharma, B., Bhatt, R.M., Srivastava, H.C., Sreehar, U. and Dash, A.P. (2011) Chlorfenapyr: a new insecticide with novel mode of action can control pyrethroid resistant malaria vectors. *Malaria Journal* 10, 16.

Satelle, D.B., Cordova, D. and Cheek, T.R. (2008) Insect ryanodine receptors: molecular targets for novel pest control chemicals. *Invertebrate Neuroscience* 8, 107–119.

Schleier III, J.J. and Peterson, R.K.D. (2011) Pyrethrins and pyrethroid insecticides. In: Oscar, L. and Jose, G.F. (eds) *Green Trends in Insect Control*. Royal Society of Chemistry, London.

Scott, C. and Bilsborrow, P. (2017) *A further investigation into the impact of the ban on neonicotinoid seed dressings on oilseed rape production in England, 2015–16*. A report from Rural Business Research and the Institute of Agri-Food Research and Innovation at Newcastle University.

Stevenson, P.C., Isman, M.B. and Belmain, S.R. (2017) A global vision of new biological control products from local uses. *Industrial Crops and Products* 110, 2–9.

Walters, K. (2016) Neonicotinoids, bees and opportunity costs for conservation. *Insect Conservation and Diversity* 9, 375–383.

Wilson, M.F. (1995) Monitoring and adapting to the changes in pesticide use profiles that occur in response to modern pest control and environmental requirements. In: Best, G.A. and Ruthen A.D. (eds) *Pesticides: Developments, Impacts and Controls*. Woodhead Publishing, London.

Wright, D.J., Peters, A., Schroer, S. and Fife, J.P. (2005) Application technology. In: Grewel, P. (ed.) *Nematodes as Biocontrol Agents*. CAB International, Wallingford, UK.

4 Herbicides

The initial development of herbicides post-World War II was introduced in Chapter 1. Following on from this, there was an extraordinary increase in the number of herbicides required for use in different crop situations. Herbicides can be classified in different ways. Selective herbicides were developed to kill weeds without damaging the crop, while non-selective herbicides kill or injure all plants within an area being treated. They are also used now where the crop has been genetically modified, so that the crop is tolerant of a non-selective herbicide. Herbicides are now often classified according to their mode of action as this helps to understand the need to choose a suitable alternative herbicide to control weeds that have become resistant to one used previously. Some herbicides can be applied *pre-planting* to kill weed seeds before the crop is sown. After sowing, a *pre-emergence* application will selectively affect weed species without interfering with the germination and growth of the crop. When rainfall is erratic and crop establishment is difficult, the farmer may prefer a *post-emergence* treatment, and this is an easy option if the crop is tolerant of the herbicide. Use of herbicides in the USA has increased less on genetically engineered crops compared with conventional crops, reflecting the ability to kill any non-crop plant in the field (Kniss, 2017).

The following description of many herbicides developed in the second half of the 20th century provides only brief details, so any user should consult specialist and official guidance on what crops can be treated and other instructions that must be followed. In the UK, the annual *UK Pesticide Guide* provides detailed information about products that are registered for use. Information on pesticides is also available on a number of websites. Overuse of a particular herbicide within an area results in weed species with resistance to the herbicide selected. In response to this, farmers and their advisers need to consult the Herbicide Resistance Action Committee (HRAC) web page (www.hracglobal.com) to obtain information on the grouping of herbicides with the same mode of action that should be avoided in seeking an alternative control for the resistant weeds (see Table 4.1). The Weed Science Society of America (WSSA) has similar information on their web page (www.wssa.net).

Non-selective Herbicides

Apart from agriculture, non-selective herbicides are used in very diverse situations, including forestry and amenity areas, alongside roads and railway tracks and in urban and industrial areas on pavements.

Paraquat

Paraquat, a bipyridylium, is a non-selective foliar contact herbicide manufactured and marketed by ICI from 1962. It was promoted as an ideal herbicide to kill both grasses and broad-leaved weeds in no-till agriculture. Spray deposits were rainfast as soon as the spray had dried and rapidly stopped further growth of the weeds, while spray reaching the soil was not active against weeds. Concern about the persistence of paraquat in soil, following repeated applications, led to detailed studies that indicated that it may be broken down only slowly, if at all, in the soil under field conditions, due to its unusual property of strong binding in the soil (Fryer *et al.*, 1975). Bromilow (2003) reviewed the worldwide use of paraquat over nearly 40 years and concluded that the cost–benefit analysis showed no deleterious side effects on non-target organisms and that paraquat, used in low tillage, is ideally compatible with the principles of sustainable agriculture. Unfortunately, it is extremely toxic to humans and other mammals leading to acute respiratory distress syndrome (ARDS) if anyone drinks the concentrated formulation. Several people died when it was decanted into unlabelled bottles and many used paraquat to commit suicide, with only two teaspoons (10 ml) sufficient to cause death. This led the manufacturer to incorporate an emetic and a stench in the formulation to deter its use as a suicide weapon. In addition, packaging was improved to prevent misuse, and in countries such as Sri Lanka the use of a locked box for pesticides was encouraged to curtail access to it, except by the farmer.

Paraquat was widely used to control weeds in rubber and oil palm plantations, with supervised teams of spray operators (Fig. 4.1). The herbicide is diluted in water using a coarse spray with droplets too large to be inhaled. Nevertheless, in the EU, paraquat has been banned since 2007, while in the USA it is classified as 'restricted use' and can only be applied by licensed spray operators.

Table 4.1. The herbicide groups as arranged by HRAC with some examples.

HRAC group	Site of action	Chemical family	An example*
A	Inhibition of acetyl CoA carboxylase (ACCase)	Aryloxphenoxy-propionate (FOPs)	fluazifop-P-butyl
		Cyclohexanedione (DIMs)	clethodim
		Phenypyrazoline	pinoxaden
B	Inhibition of acetolactate synthase ALS	Sulfonyurea	amidosulfuron
		Imidazolinone	imazmox**
		Triazolopyrimidine	florasulam
		Pyrimidinyl(thio) benzoate	
		Sulonylaminocarbonyl-triazolinone	propoxycarbazone-sodium
C1	Inhibition of photosynthesis at photo system II	Triazine	atrazine***
		Triazolinone	metamitron
		Uracil	
		Pyridazinone	
		Phenyl-carbamate	phenmedipham
C2	Inhibition of photosynthesis at system II	Urea	linuron
		Amide	propanil***
C3	Inhibition of photosynthesis at system II	Nitrile	bromoxynil
		Benzothiadiazinone	bentazon
		Phenyl-pyridazine	pyridate
D	Photosystem I-electron diversion	Bipyridylium	paraquat***
E	Inhibition of protoporphyringen oxidase	Diphenylether	fomesafen***
		Phenylpyrazole	pyralufen-ethyl
		N-phenylphthalimide	flumioxazin
		Thiadiazole	fluthiacet-methyl***
		Oxadiazole	oxadiazon***
		Triazolinone	carfentrazone-ethyl
		Oxazolidinedione	pentoxazone***
		Pyrimidindione	butafenacil***
F1	Bleaching – inhibition of carotenoid biosynthesis at the phytoene desaturase step (PDS)	Pridazinone	
		Pyridincarboxamide	diflufenican
F2	Bleaching – inhibition of 4-hydroxyphenyl-pyruvate-dioxygenase (4-HPPD)	Triketone	mesotrione
		Isoxazole	isoxaflutole
		Pyrazole	benzofenap***
F3	Bleaching – inhibition of carotenoid biosynthesis	Triazole	amitrole
		Isoxazolidinone	clomazone
		Urea	fluometuron***
		Diphenylether	aclonifen ***
G	Inhibition of EPSP synthase	Glycine	glyphosate
H	Inhibition of glutamine synthetase	Phosphinic acid	glufosinate-ammonium
I	Inhibition of DHP synthase	Carbamate	asulam

Continued

Table 4.1. Continued.

HRAC group	Site of action	Chemical family	An example*
K1	Microtubule assembly inhibition	Dinitroaniline	trifluralin***
		Pyridine	dithiopyr***
		Benzamide	tebutam***
		Benzoic acid	dacthal***
K2	Inhibition of mitosis/microtubule organization	Carbamate	chlorpropham
K3	Inhibition of cell division	Acetamide	napropamide
		Oxyacetamide	flufenacet
		Tetrazolinone	fentrazamide***
L	Inhibition of cellulose synthesis	Nitrile	dichlobenil***
		Benzamide	isoxaben
		Quinoline carboxylic acid	quinclorac***
M	Membrane disruption	Dinitrophenol	DNOC***
N	Inhibition of lipid synthesis	Thiocarbamate	triallate
		Benzofuran	ethofumesate
		Chloro-carbonic acid	dalapon***
O	Action like indole acetic acid	Phenoxy-carboxylic acid	2,4-D
		Benzoic acid	dicamba
		Pyridine carboxylic acid	clopyralid
		Quinoline carboxylic acid	quinmerac
P	Inhibition of auxin transport	Phthalamate	naptalam***
Z	Unknown	Aryaminopropionic acid	flamprop-M-methyl***

Note: *Example from *The UK Pesticide Guide*
**Used in a mixture
***Not registered in UK

Fig. 4.1. Spray team in estate in Malaysia.

Diquat

Developed in the late 1950s, diquat is a non-residual bupyridyl contact herbicide to control broad-leaved weeds and grasses. It has also been used as a pre-harvest desiccant, especially on potatoes after the use of sulfuric acid was banned. Within the EU there is a call to ban diquat by the end of 2019. It was the only herbicide to be approved in New Zealand to control unwanted target weed species in freshwater as it is rapidly removed from the water and deactivated by adsorption on sediments and various compounds in the water.

Glyphosate

This is a phosphono amino acid that, when sprayed on plants, is translocated downwards as well as into the foliage. It is non-selective but is generally more effective against grasses compared with broad-leaved weeds. It is not effective in soil. Formulations usually contain a surfactant to increase spread over foliage and to speed up penetration into leaves, an important factor in areas where rainfall can occur within two hours of a spray application. Compared with paraquat, it is slower in action so it may take ten days before weeds are obviously affected. Use of glyphosate expanded very rapidly when genetically modified crops were released that were tolerant of glyphosate, having an alternative enzyme system so plants can continue to grow without damage. Thus a farmer can delay application of glyphosate, especially in seasons where initial rainfall is erratic. The system fitted no-till or conservation in agricultural programmes and reduced the number of sprays needed, as well as requiring less fuel to apply sprays. The vast increase of glyphosate-tolerant crops – maize, cotton, soybean – has resulted in extensive areas being treated with only glyphosate to achieve weed-free crops. Inevitably, without rotation to other crops and herbicides, weeds with glyphosate resistance have been selected. Use of glyphosate has unfortunately been extended to include a pre-harvest treatment as a desiccant on cereal crops, including wheat and oats (Fig. 4.2), despite being slow-acting, as this removes late-season weeds that can be an important source of seeds for birds.

Glyphosate has also been applied using 'weed wipers', due to it being translocated from foliage downwards within a weed. Initially, the herbicide was sprayed horizontally to treat weeds that were higher than the field crop (Fig. 4.3), but the development of weed wipers enabled the offending weeds to be touched with a surface that was sufficiently wet to transfer the herbicide to the weed without dripping it on the crop. There has also been concern about the surfactants used in glyphosate formulations. These are included to increase the spread of the herbicide on leaf surfaces and its uptake by the plants. Polyoxyethylene tallow amine has been used, but is known to be toxic to aquatic organisms. Additional chemicals within a pesticide, referred to as adjuvants, have caused concern as they have not been under the same regulatory controls as the active ingredients in pesticide products (Mesnage and Antoniou, 2018).

Fig. 4.2. Tractor sprayer applying glyphosate to control late weeds prior to harvest.

Fig. 4.3. In the USA, straight jets are used to control weeds above a crop.

The International Agency for Research on Cancer (IARC), attached to the World Health Organization, reported in a monograph that they considered that glyphosate was probably carcinogenic to humans (see also Chapter 9). This is a category where there is limited evidence of carcinogenicity in humans, but sufficient evidence in experimental animals. In January 2017, the European Commission decided, in response to a European Citizen's Initiative, to call for a ban on the use of glyphosate. However, soon afterwards the European Commission stepped in with an extended approval of glyphosate until December 2017. This exemplifies the gulf between those who wish to ban all pesticides and yet do not appreciate the scientific evidence collected before a product can be registered. Laboratory animals are tested with higher rates of pesticides than used by farmers applying them in their fields. Glyphosate has been available since 1971 without any sign that it is a cause of cancers in humans.

Glyphosate has also been used in a mixture with 2,4-D as Enlist Duo, from 2014. As 2,4-D is volatile, a special formulation is used with Colex-D technology to reduce the volatility of the spray and enable farmers to control weeds resistant to glyphosate. This aroused concerns, as the public relate 2,4-D to Agent Orange, used as a military herbicide to kill vegetation in war zones. However, the illnesses associated with the exposure to Agent Orange were due to dioxin, which was a contaminant of 2,4,5-T, also present in Agent Orange. The increase in weeds resistant to glyphosate has led to the development of crops tolerant of other herbicides, notably glufosinate and dicamba.

The use of glyphosate has been reviewed by Benbrook (2016).

Glufosinate

In the 1970s, a racemic mixture of phosphinothricin was synthesized and marketed as glufosinate, a glutamine synthetase inhibitor that controlled a range of weed species. Later, in 1995, canola was the first crop genetically modified to be tolerant of glufosinate, followed by maize, cotton and soybean. The tolerance to glufosinate was achieved by inserting *bar* or *pat* genes from *Streptomyces* into these crops, so that they can detoxify phosphinothricin and prevent it doing damage to the plants. Previously, the amino acid phosphinothricin had been isolated from species of *Streptomyces*. As pointed out earlier, an alternative herbicide was needed to control glyphosate-resistant weeds, hence the significance of having glufosinate-tolerant crops.

Other broad-spectrum herbicides include monuron, a phenylurea herbicide introduced in the 1950s but now obsolete and no longer manufactured; and bromocil, introduced in 1961 and no longer permitted in the EU. It was a soil-applied uracil herbicide and used mostly in non-crop areas. Similarly, dichlobenil was introduced in 1965 for use in non-crop areas. It is moderately persistent in soils and very persistent in water, but is no longer registered for use within the EU. Imazypyr is an imidazolinone non-selective herbicide, introduced in 1985, which was very effective against grasses that were difficult to control with other herbicides (see below for other herbicides in this group). It was used on non-crop land, but this is no longer permitted in the EU. As mentioned in Chapter 1, amitrole (aminotriazole), introduced in 1953, has been considered as an

Fig. 4.4. *Striga* (witchweed).

alternative to glyphosate where weeds resistant to glyphosate occur, while picloram, a pyridine herbicide developed in 1963, is systemic and is used to control broad-leaved weeds on non-crop land.

Imazapyr

Imazapyr is a non-selective herbicide, which can control a broad range of weeds including terrestrial annual and perennial grasses and broad-leaved herbs, woody species and riparian and emergent aquatic species. It was introduced in 1985 but is no longer registered within the EU. An interesting use of it is to control witchweed (*Striga*) (Fig. 4.4), a parasitic plant (Orobanchaceae) that occurs in parts of Africa, Asia and Australia. The weed is very serious in cereal crops, with the greatest effects being in savanna agriculture in Africa. Herbicide use has been limited among poor-resource farmers who are now encouraged to adopt a 'push–pull' technique (see Chapter 8), but with new technology, one method is to grow GM hybrid maize varieties resistant to the herbicide imazapyr, which kills the *Striga* seed as it germinates (Ransom *et al.*, 2012).

Selective Herbicides

Dicamba

This plant-growth regulator was initially discovered when early development of 2,4-D was in progress but was marketed more after 1960. Like 2,4-D, its use causes vapour drift from spray deposits, so there has been reluctance to support its application on crops genetically modified to be tolerant to it and to kill weeds that are now resistant to glyphosate. The claim is that a new formulation registered in 2015 as Xtendimax, using the dicamba diglycoamine salt with VaporGrip technology, is less volatile and thus should not cause damage on susceptible crops downwind of the GM crops. Unfortunately, it seems that some farmers used older dicamba products, or the new formulation has not functioned as expected on farms. Damage has already been caused on some farms in the USA since the introduction of GM crops tolerant to dicamba.

Organic arsenical

Monosodium methanearsonate, also referred to as methylarsonic acid and known by the common name MSMA, has been used since the 1960s as a post-emergence herbicide to manage various crab-grass species as well as goosegrass and dallis grass in turf in the USA. It has also been used in some crops, in forestry and along roadsides. It was considered to be relatively safe and had a WHO class III rating until relatively recently, It is now in class II. Its use is now restricted and banned on golf courses.

Pyridine

Clopyralid was introduced in 1977 as a foliar translocated herbicide that acts as an auxin plant growth regulator, in a similar way to 2,4-D. It is used to control many broad-leaved weeds in a range of vegetable and fruit crops. It is particularly effective against thistles. Unfortunately, residues of clopyralid can persist in dead foliage and, consequentially, it has been found in composts. This led to some composts containing it causing damage to plants such as tomatoes, and led to a ban on its use on lawns in some parts of the USA. The real problem is whether the concentration of clopyralid in a compost is sufficient to affect sensitive plants. If compost is suspected to be contaminated, it should not be used in home gardens, in glasshouses or where young seedlings are being grown. It can be used on turf and in fields where the compost is ploughed in and non-legume crops are grown.

Triclopyr has been marketed since 1979, mainly as an alternative to 2,4,5-T and is used to control brushwood in rights of way and to defoliate woody plants in uncultivated areas. In the UK it has approval up to 31 October 2020. Picloram, introduced in 1963, is a similar but more persistent herbicide and is used mainly in non-crop areas.

Dithiopyr, introduced in 1991, is a pyridine that is particularly effective against crabgrass and is used on lawns and other turf areas such as golf courses and sports fields in the USA.

Aminopyralid was first registered in the USA in 2005 and was also marketed in the UK, also in mixtures with propyzamide or triclopyr, but it suffers from the same problem as clopyralid, described above, and was of such concern to vegetable growers that its use in the UK was suspended. It was subsequently reinstated in 2009 and can be used until the end of 2018, with strict control of manure, but symptoms of aminopyralid damage have since been reported. It is non-volatile but highly soluble in water. It is used only in the UK in mixtures with fluroxypyr, or with metazachlor, picloram, propyzamide or triclopyr for long-term control of noxious and invasive broad-leaved weeds in grasslands or winter-sown canola. Fluroxypyr is a newer pyridine herbicide used as a post-emergent treatment to control broad-leaved weeds. It is also marketed with a number of other herbicides, including florasulam, 2,4-D, clopyralid and dicamba.

Semicarbazone

Diflufenzopyr is a new plant-growth regulator, similar to dicamba, released in 1999, to be applied post-emergence on broad-leaved weeds and perennial grasses in pastures, no-till burndown areas and non-crop land. When applied in a mixture of dicamba, it allows a low rate of dicamba to be used against a wide range of sensitive weeds. When used to clear a field, a crop cannot be planted for 120 days.

Dinitroaniline

Trifluralin was introduced in 1961 as a pre-plant soil-incorporated or pre-emergence application to control many annual grasses and broad-leaved weeds. It acts by interrupting root development of seedlings and became one of the most widely used herbicides in the USA. In the 1960s, in Rhodesia, it was used as a pre-plant application where cotton was grown and incorporated into the soil to minimize vapour loss. In 2008, its use was banned due to its high toxicity to fish and other aquatic organisms. Among other volatile pesticides it was also detected in air samples.

Pendimethalin

Another dinitroaniline herbicide, pendimethalin, introduced in 1974, remains an important pre-emergence and early post-emergence control of grass weeds, such as black grass, in cereals and other crops. It is often formulated with another herbicide to support a different mode of action against grass weeds, where isoproturon and trifluralin are no longer registered.

Chloroacetamides

In 1964, propachlor was the first chloroacetamide to be developed as a pre-emergent herbicide to control annual grasses and broad-leaved weeds. It was followed in 1966 by alachlor, which has been widely used in the USA. In 1974, metolachlor, which was a mixture of (S) and (R) stereoisomers was introduced, but since 1996 the (S) isomer has replaced it. In 1985, another chloroacetamide, acetochlor, was commercialized, but together with propachlor and alachlor is not registered in the EU. Lastly, in 1999, dimethenamid-P was developed but is only used in mixtures in the UK, e.g. with metazachlor (a newer chloroacetamide introduced in 1982), quinmerac (a quinoline introduced in 1993) or both. These are ergosterol inhibitors of seedling growth.

Carbamate

Asulam, developed in 1965, is a translocated carbamate herbicide that has been used principally to control bracken, particularly on grazing pastures, moorland and amenity grassland. It has been applied aerially in some areas for which special permits are required, and in 2017 an emergency authorization of its use was allowed to control invasive bracken for 120 days. It is no longer registered by the EU due to concerns about bioaccumulation, but it is hoped that the EU will be persuaded to allow use on bracken to continue.

Another carbamate, introduced in 1951, is chlorpropham, a residual herbicide to control annual and perennial broad-leaved weeds in various crops. It is also used as a potato sprout suppressant and is applied as a fog to percolate through large quantities of potatoes in stores.

Prosulfocarb is a thiocarbamate introduced in 1988 to control annual grasses and broad-leaved weeds in a range of crops, usually cereals and potatoes, extended to certain other crops by the extension of authorization for minor uses.

Arylalanine

In 1978, flamprop-M-methyl, a translocated, post-emergence herbicide was marketed for controlling wild oats in wheat. It is not registered in the EU.

Sulfonylureas

The first sulfonylurea herbicide was discovered in 1975 and introduced a unique mode of action, by inhibiting the synthesis of amino acids, such as valine, isoleucine and leucine, and aceto-lactase inhibitors (ALS). Young weeds starve as they cannot produce proteins needed for growth. Crops such as rice, wheat, barley, soybean and maize can metabolyse sulfonylureas safely, so this group of herbicides has been used extensively and is applied at very low doses of active ingredient. This required a change for farmers as they were used to applying herbicides in much larger doses and not in g/ha. They now form one of the largest groups of herbicides with many different actives registered across the world.

Chlorsulfuron was first registered in 1980 and used the following year on small-grain crops. Subsequently, other similar herbicides were developed including metsulfuron (1983), primisulfuron (1987), rimsulfuron

(1989), nicosulfuron (1990), amidosulfuron (1992), sulfosulfuron (1995), iodosulfuron (1999) and mesosulfuron (2001). Some of these are used as mixtures, thus iodosulfuron and mesosulfuron are registered for use in the UK as a mixture on winter wheat, with an extension of use on some similar crops, such as rye.

The different sulfonylureas offer weed control as a pre- or post-emergence treatment, with choice of length of residual control. They are not volatile, so there is no off-target movement from spray deposits.

Another herbicide with a similar mode of action is propoxycarbazone-sodium, a triazolone introduced around 2000 as a post-emergent treatment to control blackgrass in winter wheat. In the UK, the maximum total dose is one full-dose treatment and not in a programme where other ALS inhibitors are used. Thiencarbazone-methyl is another new herbicide that is often used in mixtures with isoxaflutole for pre-emergence control of grassy and broad-leaved weeds in maize, soybeans, wheat, turf and ornamentals. A product with the mixture is marketed and claimed to be reactivated following rainfall to control late weeds.

Other ALS inhibitors are the imidazolinone herbicides. Imazamethabenz was recorded in 1982, followed by imazaquin (1983), imazethapyr (1984), imazamox (1995) and imazapic (1997). Some are used as selective herbicides, thus imazamethabenz provides good control of wild oats and volunteer canola in spring-sown barley, but it is not registered within the EU. Imazaquin is used to control weeds in grass and turf. Imazathapyr controls grasses in legume crops. Imazamox is used to control reeds and grasses in aquatic areas but is also used with metazachlor to control broad-leaved weeds and annual grasses in canola or with pendimethalin in legume crops, respectively. Imazapic is also selective and can be used in legume crops, but is not registered in the EU.

Propinionates

Dicofop-methyl, introduced in 1975, is applied as a post-emergence herbicide with MCPA and mecoprop-P to control annual grasses including wild oats in wheat, and is also used with ferrous sulfate and MCPA in managed amenity turf production. Fluazifop-P-butyl was first marketed in 1981 and applied as a post-emergence herbicide to control annual grasses including wild oats in wheat and broad-leaved crops and ornamentals. Quizalofop-P-ethyl, introduced in 1989, is another post-emergence herbicide used for grass weed control alongside quizalofop-P-tefuryl as an alternative to others with the same mode of action. Fenoxaprop-P-ethyl, introduced in 1989, is a post-emergence herbicide to control annual and perennial grasses and has been used in rice crops. In 1990, clodinafop-propagyl was first marketed as an alternative propinionate post-emergent herbicide to control annual grasses and can be applied mixed with cloquintocet-mexyl, which is a herbicide safener that accelerates the herbicide detoxification in the treated crop. Prosulocarb, or pinoxaden, another post-emergence herbicide, was introduced in 2006.

Nitrophenyl ether

Acifluorfen-sodium was introduced in 1980 and is applied pre-emergence to soybeans, other legumes and rice, but is not used in the EU. It controls annual broad-leaved weeds. A complex ester of aciflorfen is known as lactofen, introduced in 1987. Fomesafen is a similar herbicide, also used on potatoes, tomatoes and cotton. It can be used pre-plant up to 14 days before planting and provides weed control for up to six weeks in genetically modified and conventional cotton, with control of glyphosate-resistant Palmer pigweed. Any post-emergence treatment needs to be with shielded precision equipment to avoid spray on the stem or foliage of the cotton plants. Fomesafen is more effective than fluometuron, which had been used on cotton since the 1960s. Oxyfluorfen, introduced in 1976, is used in some EU countries as a pre- or post-emergence treatment to control annual weeds in various vegetable, fruit and other crops. It has also been used in sugar cane and non-crop areas.

N-phenyl phthalimides

Flumioxazin is for pre-emergence broad-spectrum control of weeds near or in water to control algae and pond weeds. It was introduced in 1994, and in the UK it is registered for one early post-emergence treatment per crop in winter wheat and oats, provided plants have been hardened by cool weather and are not lush with soft growth. Flumiclorac-pentyl, released in 1995, is a dicarboximide, which has been used to control certain problematic weeds.

A number of other herbicides with different chemical structures were introduced in the second half of the 20th century. Fluometuron (a phenylurea) has a similar mode of action as mesotrione and was developed in 1964. It was used as a soil-applied selective herbicide to control annual grasses and broad-leaved weeds in cotton and sugar cane. Oxadiazon, an oxadiazole, is a pre- and early post-emergent herbicide developed in 1969 and is used to control bindweed and many annual broad-leaved weeds. It acts by inhibiting protoporphyrinogen oxidase (PPO). It is practically insoluble in water. Tebutam, a benzamide, was reported in 1976 as a pre-emergence herbicide used to control broad-leaved weeds and grasses in brassicas and other crops. It is not registered in the EU. A nitrophenyl ether, aclonifen is used for the pre-emergence control of grass and broad-leaved weeds. It was introduced in 1983 and used on groundnuts, beans and soybeans in many EU countries (but not the UK) and also in rice. In 1985, clomazone, an isoxazolidinone, was introduced. It acts by inhibiting synthesis of chlorophyll pigments. It is a residual herbicide applied to control annual broad-leaved weeds in canola, field beans and peas, as well as many crops in the UK, through extension of authorization for minor crops. One application per crop is permitted. It is also supplied in mixtures with linuron, metazachlor, metribuzin, napropamide or pendimethalin. The last of these mixtures has to be applied with a coarse spray.

Later, in 1987, a cyclohexanedione herbicide, clethodim, was introduced as a post-emergence grass herbicide to control weeds in certain broad-leaved crops such as cabbages, beans, peas and carrots. There is a requirement of a 14-day gap before and after application. Also in 1987, benzofenap, a benzoylpyrazole, was registered for controlling broad-leaved weeds in water-seeded rice in Asia. It is not registered in the EU. Quinclorac, a quinolinecarboxylic acid, was produced in 1989 as a post-emergence treatment to control broad-leaved weeds in rice and grass. It is not registered in the EU. In 1991, a triazolinone, sulfentrazone, was introduced for broad-leaved weeds, but it can also control some grasses such as sedges in turf. It is highly soluble in water and volatile. It has not been approved within the EU.

Carfentrazone-ethyl, a phenylpyrazoline developed in 1997, is also primarily for control of broad-leaved weeds. In the UK, it is used on cereal and potato crops. It is also available as a mixture with mecoprop-P. Also that year, fentrazamide, a tetrazolinone, was introduced as a post-emergence herbicide to control barnyard grass, annual sedges and broad-leaved weeds in rice crops, while pentoxazone, an oxazolidinedione, was developed as a pre- and post-emergence herbicide in rice in 1997. Neither of these two herbicides has been registered in the EU. In 1999, a thiadiazole, fluthiacet-methyl, was introduced and has been used to control broad-leaved weeds in cotton, maize and soybeans in the USA. It has been effective against some of the difficult-to-control weeds resistant to glyphosphate.

Several New Herbicides Introduced since 2000

Butafenacil, a pyrimidindione, was introduced in 2000 to control annual and perennial broad-leaved weeds in fruit and other crops. It is not registered in the EU. Also propoxycarbazone-sodium, a triazolone, was introduced in 2000 as a new residual grass weed herbicide aimed at controlling black grass. A triketone, mesotrione, introduced in 2001, is a pre- and post-emergent herbicide used mostly in maize crops to control some grass and broad-leaved weeds. It acts as a 4-hydroxyphenylpyruvate dioxygenase inhibitor. In the UK, it is also registered for use in a mixture with nicosulfuron. Topramezone, a benzoyl pyrazole, introduced in 2006, and tembotrione, introduced in 2007, have the same mode of action and provide similar control to mesotrione, but are not registered in the UK. A benzoylpyrazole, topranezone, released in 2006, is a post-emergence herbicide for broad-leaved weeds and grasses used mainly on maize. According to Grossman and Ehrhardt (2007), the tolerance of maize to topramezone is due to more rapid metabolism combined with a lower sensitivity of the 4-HPPD target enzyme. Since 2006, pinoxaden is a new phenylpyrazoline post-emergence herbicide to control grass weeds in cereals and is used in turf management. It is also marketed in mixtures with clodinafop-propagyl and florasulam.

No-till and Herbicide Use

Ploughing fields goes back millennia and was used in south-east Europe 8000 years ago, with the Romans using an iron plowshare. The area ploughed expanded rapidly only with the invention of the 'steam horse' at the beginning of the 20th century and led, in the USA, to severe erosion culminating with the dust bowl in the 1930s. Conservation tillage began after World War II with the availability of 2,4-D and now covers 95 million ha globally (Lal *et al.*, 2007). No-till farming is defined as a method of growing crops or pasture from year to year without disturbing the soil through ploughing. Shallow tillage may be needed to provide a surface that allows seeds to germinate. No-till can reduce or eliminate soil erosion, caused by rain washing away the loosened soil. There is considerable evidence that ploughing can have an adverse effect on some soil-dwelling organisms, so a no-till programme is considered environmentally better (Fig. 4.5). Survival of earthworms and other organisms is better, and this allows increased organic matter and water retention in the soil, using crop residues as a mulch. However, it is not suitable on all soil types, such as poorly drained clay soils. As ploughing buried most weed seeds, in the absence of ploughing, weed management does necessitate applying herbicides. With paraquat, it was possible to wait for the first rainfall to enable weed seeds to germinate and then spray to kill the weeds before sowing the crop.

Fig. 4.5. Direct drilling. (Photo courtesy of Santiago del Solar Dorrego, used with permission).

The genetic manipulation of a crop to be herbicide-tolerant was welcomed in many countries, but so far there has been too much reliance on using one herbicide, glyphosate, to control the weeds. Some new cultivars tolerant to 2,4-D and dicamba are becoming available but this depends on suitable formulation of the herbicides to avoid vapour drift. Clearly, longer-term rotation systems, both for crop and herbicide use, need to be considered to avoid the situation that has developed where several different glyphosate-tolerant crops are grown in one locality.

Precision Farming

The availability of GPS on tractors has introduced the concept of only applying a herbicide where weeds are present in fields. Initially, this was examined by walking fields to determine where there were distinct patches of weeds. A recent refinement of this is to use a drone, equipped with a suitable camera to detect the presence of weeds. The sprayer could then be set to spray those areas as it passed across the identified patch. Subsequently, there has been much attention given to try to detect weeds as the sprayer passes through a field so that individual weeds can be spot-treated.

In Australia, farmers have started to adopt the use of equipment that crushes weed seeds in the trash that is left by a combine harvester, to minimize the amount of weed seed left on the soil surface. It is considered that this affects weed seeds that are most likely to have some resistance to the herbicides being used.

References

Benbrook, C.M. (2016) Trends in glyphosate herbicide use in the United States and globally. *Environmental Services Europe* 28.

Bromilow, R.H. (2003) Paraquat and sustainable agriculture. *Pest Management Science* 60, 340–349.

Fryer, J.D., Hance, R.J. and Ludwig, J.W. (1975) Long-term persistence of paraquat in a sandy loam soil. *Weed Research* 15, 189–194.

Grossmann, K. and Ehrhardt, T. (2007) On the mechanism of action and selectivity of the corn herbicide topramezone: a new inhibitor of 4-hydroxyphenylpyruvate dioxygenase. *Pest Management Science* 63, 429–439.

Kniss, A.R. (2017) Long term trends in the intensity and relative toxicity of herbicide use. *Nature Communications* 8.

Lal, R., Reicosky, D.C. and Hanson, J.D. (2007) Evolution of the plow over 10,000 years and the rationale for no-till farming. *Soil & Tillage Research* 93, 1–12.

Mesnage, R. and Antoniou, M.N. (2018) Ignoring adjuvant toxicity falsifies the safety profile of commercial pesticides. *Frontiers in Public Health* 5, 361. DOI: 10.3389/fpubh.2017.00361.

Ransom, J., Kanampiu, F., Gressel, J., De Groote, H., Burnet, M. and Odhiambo, G. (2012) Herbicide applied to Imidazolinone resistant-maize seed as a *Striga* control option for small scale African farmers. *Weed Science* 60, 283–289.

5 Fungicides

As discussed earlier, man has been concerned about plants succumbing to diseases as far back as ancient Greek and biblical times, when sulfur was known as brimstone or 'burning stone'. The Ebers Papyrus, dating from 1550 BC, describes an eye salve containing sulfur. Ancient priests thought hell consisted not only of fire but also of burning brimstone. The Greeks burnt sulfur to purify their temples. The ancients recognized the pungent odour and that rats were killed when exposed to sulfur dioxide vapours. This introduced the idea of using burning sulfur as a fumigant to kill pests. In *The Odyssey* Homer talks about burning sulfur to preserve corpses in the hot sun.

Fungicide development can be divided into roughly three eras: the inorganic period (including organo-metallics); the synthetic organic protectants; and the organic systemic fungicides. However, there are significant overlaps and exceptions, notably that some of the most modern chemicals are not very mobile in plants, and those that are, mostly move up the plant in the xylem stream and not downwards. There has been a substantial increase in intrinsic fungitoxicity over the past 100 years, with a corresponding reduction in dose rates. A further general trend is that most of the older chemicals used have a multi-site mode of action (in fact, discerning their modes of action defied researchers for a long period), whereas more modern fungicides tend to have a single active site mode of action.

Fungicide discovery started by testing chemicals known to be toxic to other forms of life, for example materials based on copper and arsenic. Later compounds were often by-products from other chemical processes, spurred by the desirability of finding new uses for what might otherwise be waste products. The dithiocarbamates, for example, were by-products from the vulcanization of rubber. From this, it is a small step to the establishment of screens for biological activity, and a significant number of chemical and oil companies took this route after World War II. Screening programmes in the 1960s and 1970s became increasingly large-scale (and costly); fungicidal activity being sought alongside herbicidal, insecticidal or rodenticidal activity; indeed any form of biological activity that might have a commercial outcome. More recently, biorational design has come to the fore, specific compounds being sought that should interact with target metabolic steps in fungi of interest, based on knowledge of their biochemistry. By such means, classes of molecules with likely fungitoxicity can be predicted, thus moving away from the hugely expensive random screening approach.

This all assumes that one knows how to detect fungicidal activity in the first place: something much more challenging than it might seem at first glance. Early bioassays were designed to measure inhibition of spore germination or of mycelial growth *in vitro*. The former, usually conducted on glass microscope slides or plastic micro-beakers in the presence of a trace deposit of the chemical to be tested, had the advantage of being sufficiently small that a large number of replicates and dose rates could be tested, and the results lent themselves to statistical analysis. However, the many variables involved, not least the choice of target organism, meant that comparisons between laboratories were difficult until a standardized design was adopted, notably that proposed by the American Phytopathological Society in 1943. Inhibition of mycelial growth can be assessed by incorporating a range of doses of the chemical being tested into a nutrient agar. This approach is open to many criticisms; for example, the test compound might be complexed by components of the agar medium, and rate of spread across agar is not necessarily a good measure of the true growth rate of a fungal colony. A more recent approach to measuring growth inhibition is to use liquid culture in microtitre plates and to assess growth by means of light absorption.

By the 1960s, there was a growing realization in the industry that testing fungitoxicity *in vitro* had huge drawbacks: many promising chemicals failed later development because of instability or toxicity towards non-target organisms, and there was growing evidence that large numbers of possible crop protectants were being missed by *in vitro* screens; for example, no chemical that worked by inducing resistance in a plant could possibly be discovered in this way. Despite the costs, industry moved to conducting primary screening on plants, miniaturized as far as possible by using seedlings, detached leaves and fast-growing or dwarf relatives of crop plants. Nevertheless, the scale of operation was still considerable given that each candidate chemical would most likely be tested at two different concentrations, would be applied in two different ways (foliar spray and root drench), and assayed against up to seven different diseases. When replication and the need for safety precautions are included, clearly for a throughput of, say, 200 chemicals per week, a dedicated and highly automated glasshouse facility would be needed. Testing on this basis facilitates the identification of promising compounds and provides much useful ancillary information, for example on herbicidal or growth-regulating activity.

It is worth noticing in passing that crop protection is not the only outlet for candidate fungicides. Chemicals that fail on plants because of phytotoxicity might still be useful in the protection of organic fabrics, wool, cotton, leather etc., and in the vast wood preservation industry. Pharmaceutical companies are another interested party, given the expanding market for antifungals in human and veterinary medicine. A quirk of nomenclature is worth noting. Although the term fungicide is widely used and accepted, many of the products so named do not actually kill fungi. What is necessary for them to work in the field is for them to protect plants, which they may do in a variety of ways that are not strictly fungicidal; for example, by stimulating plant resistance or suppressing pathogen sporulation. A further irony is that one of the original targets for fungicide use, *Phytophthora* species, are now known to be unrelated to true fungi in evolutionary terms. To plant pathologists, however, they remain as 'honorary fungi', due to their mycelial habit, their ecological behaviour and their tendency to plant pathogenesis.

The first fungicides to be developed that were based on heavy metals (copper and mercury) or sulfur, mostly as inorganic compounds or, in some cases, bound to an organic moiety to increase efficacy, were considered in Chapter 1.

Organic Protectant Fungicides

Although the fungicidal activity of the dithiocarbamates was first reported in 1934, commercial development of these and other groups of organic protectants did not occur until after World War II. Ferbam, the ferric salt of dithiocarbamic acid, was among the first of this family of chemicals to be used commercially, but it was unpopular because it left a black deposit on leaves. Thiram (tetramethylthiuram disulfide) has proved much more resilient. It has a long period of use as a seed dressing effective against Pythium and a range of other pathogens. Other members of the family include the alkyldiamines generally used as a salt formulation with zinc (Zineb), manganese (Maneb) or a complex of the two (Mancozeb). These have proved to be extremely competent protective fungicides with low-enough phytotoxicity for foliar use active against a range of diseases with the main exception of powdery mildews. Mancozeb was first reported in 1961 and remains registered within the EU. It is used to control a wide range of pathogens including blights and scab on crops, such as potatoes, tomatoes and ornamentals.

The quinones chloranil and dichlone, although actively fungicidal, had the limitation of being unstable in light. They were incorporated into soil for use against root infections of vegetable crops but rapidly became outmoded. A further member, dithianon, was introduced in 1963 and was used to control foliar pathogens on a range of crops. A much more successful group was the phthalimides, introduced as captan, the closely related folpet, in 1952, and captafol ten years later. The former was used extensively to control apple scab disease and gained the reputation as the first reliable protectant fungicide that did not blemish fruit. Other pathogens controlled on a range of vegetables and ornamentals include *Botrytis cinerea*, *Colletotrichum* spp., *Ascochyta* spp. and *Thielaviopsis basicola*. Phthalimides have a broad spectrum of efficacy with the notable exception of the powdery mildews. Diflochluanid, a related compound, has mostly been used against grey mould on a range of fruit crops, as was dichloran, an unrelated nitroaniline compound.

Substituted nitrobenzenes have had considerable use as fungicides. The chlorinated derivatives tecnazine and quintozene were introduced in the late 1940s, initially against *B. cinerea* and *Rhizoctonia solani*. When the former was used to control dry rot of potato caused by *Fusarium caeruleum*, it was found to inhibit sprouting and was used primarily for this purpose. They are unusual fungicides in a number of respects, being mainly fungistatic rather than fungicidal, and under laboratory conditions target fungi rapidly evolve resistance to them, although this does not happen to any great extent in the field. The related aromatic hydrocarbons chloroneb, 2-phenyl phenol and biphenyl are too volatile for field use but found a role in the post-harvest environment to control storage rots and moulds on fruit where volatility would be an advantage. Use was limited because of a chemical taint imparted to some fruit, and the current trend is to avoid any post-harvest chemical treatment altogether, if possible. The dinitrophenolic compounds dinocap and binapacryl are some of the few from this era that gave good control of powdery mildews. Binapacryl was for many years a staple component of orchard spray mixes, together with captan, to control *Podosphaera leucotricha* on pome fruit.

An unusual fungicide that found use against apple and pear scab and cherry leaf spot was dodine (N-dodecylguanidinium acetate), patented in 1959 and used under the trade name Cyprex. As well as protecting against infection it had significant eradicant and curative action and could be applied after an infection period in the knowledge that it could prevent an established infection from developing. Being a cationic surfactant, it readily redistributed on leaf surfaces that could compensate for poor initial coverage. A final member of the protectant fungicide group is chlorothalonil. Although introduced in 1964 it continues to be used, mainly in mixtures with other compounds, since fungi seem to have had little success in developing resistance to it.

Systemic Fungicides

The holy grail of fungicide research in the mid-20th century was to discover fungicides that moved within plants to have both a protective and curative effect. Experience with herbicides, many of which are mobile in plants, and with the organophosphate insecticides suggested that this was an attainable goal. Different methods of screening compounds for biological activity (see above) were part of the key to success. Along the way there had been some partial successes. The antibiotic griseofulvin, discovered in the 1940s, moved systemically in plants and gave control of some diseases, but was too unstable for general use. Another antibiotic, cycloheximide (Actidione), was also mobile in plants and was used experimentally against white pine blister rust (caused by *Cronartium ribicola*). Approximately 4 million trees were treated in 1959. Enthusiasm surrounding promising disease control was tempered by unacceptable phytotoxicity.

Several companies reported the development of promising systemic fungicides during the late 1960s: DuPont with benomyl (1968), ICI with dimethirimol (1968), BASF with dodemorph (1967) and Uniroyal with carboxin (1966). All raised considerable interest at the first International Congress of Plant Pathology in 1968, where there was a strong feeling of optimism that the industry was entering a new phase in the fungicidal treatment of plant diseases.

Benomyl was the first of the benzimidazole group of fungicides with both protective and eradicant activity against a range of diseases of cereals, orchard fruit, vegetables and vines, but it was toxic to micro-organisms and invertebrates, especially earthworms. In aqueous media, it hydrolyses to methyl benzimidazole carbamate, thought to be the active ingredient and sold separately under the name carbendazim. Other members of the group include: fuberidazole; thiabendazole, which was originally sold as an antihelmintic before being used in crop protection, mainly post-harvest, on fruits and tubers and remains registered principally to protect seed potatoes in storage; and thiophanate-methyl, which also undergoes a chemical rearrangement in aqueous solution to give carbendazim. The unique properties of benzimidazoles had not been seen before in the protectants. These included low use rates, a broad spectrum of activity and systemic movement with post-infection action that allowed the interval between sprays to be extended. The main drawback was that many of the pathogens developed resistance to the benzimidazoles, probably because they are single-site inhibitors of fungal microtubule assembly during mitosis, via tubulin–benzimidazole interactions. The rapid decline in effectiveness against certain diseases – for example *Cercospora* leaf spot on sugar beet and eyespot disease of wheat – was a foretaste of the problems of fungicide resistance that the industry has had to live with ever since.

Dimethirimol was the first member of the hydroxypyrimidine fungicides, the others being ethirimol, used on cereals, and bupirimate, used on fruit crops and ornamentals. Their spectrum of activity is against powdery mildews, and like the benzimidazoles, they have suffered from the development of resistance in target fungi, so much so that dimethirimol was rapidly rendered obsolete and was withdrawn. Ethirimol and bupirimate also encountered resistance build-up but not to the extent of rendering them useless. The difference probably lies in the contrasting ways in which the fungicides were used. Dimethirimol was used in the closed environment of glasshouses against cucurbit powdery mildew, where its huge advantages over previous chemicals led growers to move over to it exclusively, leading to huge selection pressure in favour of any resistant mutants. Ethirimol, on the other hand, was mainly used as a seed dressing on cereals against *Blumeria graminis*, but, due to degradation and dilution as the plants grew, was only exposed to the pathogen for part of the growing season, other chemicals being used as mid-season sprays if needed. Further, not all growers adopted it, so in any one season a significant portion of the pathogen population would not be exposed to the chemical. Consequently, the selection pressure was less intense, and ethirimol enjoyed a longer period of use, eventually being replaced because better chemicals became available.

Carboxin was the first member of the carboxamides (originally referred to as oxathiins), others being oxycarboxim, fenfuram, benodanil and mepronil. Their efficacy is largely restricted to basidiomycete fungi. They have seen use as foliar sprays against rust diseases, but mostly they have been used as seed dressings effective against both loose and covered smut as well as root infections caused by *Rhizoctonia*. They did much to supplant organo-mercury as a seed dressing on cereals due to their penetrative effect against loose smut (*Ustilago* spp.), which overwinters in the seed embryo and is unaffected by mercury. Their mode of action seems to be inhibition of succinic dehydrogenase of complex II in the mitochondrial electron transport chain, although it is not clear why basidiomycetes are so much more susceptible than other fungi.

Dodemorph was the forerunner of the morpholine group of fungicides. It had little commercial success but was followed by tridemorph, fenpropimorph and fenpropidin. The spectrum of activity is strongest against powdery mildews, but the latter two, in particular, control a range of rust and other foliar pathogens. Fenpropimorph, when introduced, was effective at very low dose rates and there was a feeling in the 1980s that it was as good a foliar fungicide for cereals as was ever likely to be found. Such optimism has been tempered more recently by the evolution of reduced sensitivity in target pathogens, which is considered further later.

A feature notably lacking in the four groups of chemicals just described is activity against oomycete pathogens. This was remedied in the mid-1970s with the introduction of the first phenylamide (originally called acylalanine) fungicides, metalaxyl and furalaxyl, followed in the 1980s by benalaxyl, ofurace and oxadixyl. Members of the group are generally effective against species of *Pythium*, *Phytophthora* and the downy mildews, but with very little activity against other plant pathogens. A major use of metalaxyl was for control of potato blight where its systemic movement in the foliage was a considerable advance over the protection given by dithiocarbamates or Bordeaux mixture. This group has also suffered from resistance problems. Mode of action involves inhibition of synthesis of ribosomal RNA.

Other oomycete-active compounds introduced around the same time include propamocarb (1978), a translocated carbamate used off-label in the UK on certain crops and in forest nurseries; cymoxanil, a cyanoacetamide oxime registered in the UK specifically for use on potatoes, hops and grapes; and fosetyl (1977), a simple phosphonate compound generally used as the aluminium salt. Somewhat later, in 1988, dimethomorph was released. It is quite distinct from other morpholines with its activity against oomycetes via the inhibition of cell wall formation. It has been used for the control of late blight on potato.

The mid-1970s also saw the first triazole, imidazole, piperidine and pyrimidine fungicides, chemicals that share a mode of action, namely inhibition of a demethylation (DMI) step during the biosynthesis of fungal sterols. Many members of this 'family' have enjoyed considerable commercial success and the number of compounds increased considerably over the ensuing years. Triadimefon, a triazole, was released in 1976. It provided curative as well as protectant action at low dose rates and distributed well through the sprayed crop. It is no longer registered in the EU, but it stimulated a search for more active triazoles. A year later, imazalil, an imidazole, became available to control a wide range of fungi including *Tilletia* (a smut fungus) and *Helminthosporium* spp., causing leaf blight on fruit, vegetables and ornamentals. It has also been employed as a post-harvest dip for bananas and citrus. It has been registered throughout the EU but is no longer included in the *UK Pesticide Guide* as it is being phased out in the UK. Also in 1977, prochloraz, an imidazole, was introduced and is still registered in the EU. In the UK it is used in a mixture with propiconazole or tebuconazole (or both tebuconazole and proquinazid) to control anthracnose, dothiorella complex, stem-end rot and eyespot on cereals and other crops. A year later, triadimenol, another triazole, was introduced, mainly as a seed dressing, but it is no longer registered in the UK.

In 1979, propiconazole, another triazole, was introduced with a broad range of activity against pathogens including *Blumeria graminis*, *Leptosphaeria nodorum*, *Pseudocerosporella herpotrichoides*, (*Tapesia* spp.), *Puccinia* spp., *Pyrenophora teres*, *Rhynchosporium secalis* and the septoria diseases of small-grain cereals, enabling it to be used in a wide range of agricultural crops. This was followed by bitertanol in 1980, which was active against a range of diseases including scab, powdery mildew, rusts and blackspot, but this is no longer approved within the EU. Flutriafol appeared in 1981 and is still registered within the EU. It is a curative and preventative triazole fungicide used to control leaf and ear diseases, usually in cereals, but has been used on apples and other crops. This was followed by penconazole in 1983. This is registered in the EU for application on a wide range of crops. In 1984, fluzilazole was reported and was followed by hexaconazole, an imidazole, used to control both seed-borne and soil-borne diseases, especially those caused by ascomycete and basidiomycete pathogens. Neither of these azoles is registered within the EU.

In 1986, triflumizole, an imidazole, was registered in Japan for control of fungal diseases on top fruit, grapes and other crops. Another azole was diniconazole, but neither it nor the pyridine pyrifenox were registered in the EU. Fenpropidin, a piperidine (morpholine), released in 1986, has already been mentioned and is a key systemic, curative fungicide used to treat cereals within the EU, although in the UK it is only available mixed with difenoconazole, or procloraz and tebuconazole, to control brown and yellow rusts, eyespot, septoria glume and leaf blotch, powdery mildew and ear diseases in wheat. The first report of tebuconazole was in 1988. It has become one of the most important azole fungicides and is also marketed mixed with many other fungicides. These are two examples of the trend to use only fungicides in mixtures, preferably with others of a different mode of action.

In 1989, other broad-spectrum azole fungicides were introduced including: cyproconazole, first used in France and Switzerland but now registered within the EU; myclobutanil, used on grapes but not registered in the UK currently; and difenoconazole, which is also available mixed with azoxystrobin in the UK; it can also be mixed with carbendazim for improved control of some infections on oilseed rape.

Tetraconazole arrived in 1990, and was used to control a range of fungal infections on sugar beet, but is currently not registered in the UK. This was followed by fenbuconazole in 1992, which is registered in the UK for use on top fruit and grapevines. In 1993, epoxyconazole was first registered and is used to protect sugar beet; cereal crops including wheat, barley, rye, oats and triticale; coffee; and bananas. In the UK, registration currently runs until 2021. Triticonazole was also introduced at that point, and a year later metconazole was registered for fungal infections on fruit and other crops. It was also registered in the UK for treating cereals, certain vegetables and oilseed rape. Fluquinconazole came in 1995 as a selective protectant and curative fungicide used to control various endophytic diseases, mainly on cereals, but registration was due to end in 2017. The latest triazoles are prothioconazole and mefentrifluconazole. Prothioconazole was introduced in 2002 and is marketed in the UK in mixtures with other fungicides. It is used as a seed treatment to control seed-borne diseases and to improve establishment of the crop.

It will be apparent that the sterol demethylation inhibitors (DMI), dominated by triazoles and imidazoles, have been a fertile area for fungicide research; the list of compounds that have been introduced over the years is extensive and reaches beyond those mentioned here. Azoles have also proved useful in wood preservation and have found use as antifungals in human and veterinary medicine. Miconazole is available for topical application to eradicate fungal infections of skin and nails, while ketoconazole is incorporated into shampoo and can be taken orally to combat internal infections.

During the 1990s, another group of fungicides, the stobilurins, was marketed with a broad spectrum of activity against diseases as a respiration inhibitor (QoL fungicide) and has become the second largest group of fungicides. The discovery of this type of fungicide was inspired by a study of natural benzothiazole methoxyacrylates, ironically found in the fruit caps of certain basidiomycete fungi, the structure of which provided a starting point for their development. In 1992, azoxystrobin was the first to be registered within the EU. As a translaminar systemic protectant it is registered in the UK for a very wide range of crops as well a large number of extensions of authorization on minor crops. It is also marketed as a mixture with several other fungicides; thus with fluazinam it is used on potatoes for blight control.

Kresoxim-methyl was introduced in 1996, and in the UK it is used mainly on apples (to control scab) and on many other fruit crops. A maximum of four treatments per year is recommended on apples, with up to three on other crops. In 1998, famoxadone was introduced and fenamidone in 2002. Both are effective in controlling oomycete pathogens and are only available in mixtures in the UK. In 1999, trifluoxystrobin was added to the strobilurin group, and is often marketed in a mixture with an azole fungicide; followed by pyraclostrobin in 2000; cyazofamid (a cyanoimidazole) in 2001, which, like famoxadone, is used mainly to control blight on potatoes; and picoxystrobin and fluoxastrobin in 2002. Picoxystrobin has both vapour phase and systemic activity in cereal leaves and is effective against a wide range of diseases.

Over the period of development of azoles and strobilurins, a number of other fungicides with different modes of action were marketed. These included the dicarboximides (iprodione, vinclozolin and procymidione in the 1970s), the phenylpyrroles (fenpicolonil, fludioxonil in 1990), and benthiavalicarb-isopropyl and mandipropamid since 2000. A novel pyrazole, introduced in 2008, isopyrazam, which inhibits respiration and spore germination, has been marketed in the EU to control a wide range of diseases. In 1990, a dinitroaniline fungicide, fluazinam, was introduced to control late blight, white mould, clubroot, downy mildew, scab and *Alternaria* blotch; it also has activity controlling mites.

Of these, only a few are registered in the UK and the dicarboximides is the group that has seen the greatest use. Their mode of action is poorly understood: treated fungi exhibit hyphal swelling and repeated branching and become osmotically sensitive and susceptible to bursting of hyphal tips. They are used for the control of *Sclerotinia*, *Botrytis*, *Monilinia*, *Alternaria*, *Phoma* and *Septoria* in grapevine, oilseed rape, hops, ornamentals, fruit and vegetables, becoming increasingly important as these pathogens developed resistance to the benzimidazole group of compounds.

Organophosphates have seen limited use as fungicides, examples being triamiphos (1960) and ditalimfos (1966), available primarily as protectants against powdery mildew on ornamentals and fruit. They were superseded by better products and are no longer available. A further group, the anilinopyrimidines, was introduced in the 1990s. Their mode of action, which may involve methionine metabolism, is usefully distinct from and shows no cross-resistance with other groups. Another product, zoxamide, a benzamide (2001), is only registered in the UK when used in a mixture on various vegetable crops including tomatoes.

The overwhelming majority of compounds mentioned above are products of synthetic chemistry. Prompted by the success of antibiotics in medicine, researchers have sought natural products, typically from microbial fermentation, with fungicidal activity. Mention has already been made of griseofulvin and cycloheximide, which never entered commercial practice, but a number of others have done so, all being products of various *Streptomycetes*. The blasticidins were developed in Japan for use against rice blast disease. They were introduced in 1962, and by 1968 some 8000 tons of the crude material were used for this purpose. A rather better product, kasugamycin, was introduced in 1965 against the same target, and up to 20,000 tons were used annually until resistance was detected in 1971. It is a protein synthesis inhibitor that operates by preventing the binding of tRNA to ribosomes. The polyoxins, an antibiotic family containing a pyrimidine nucleus, were also developed in Japan and introduced in 1968, as was validamycin (1972). Both of these have been used to control rice sheath blight caused by *Rhizoctonia solani*. The polyoxins interfere with cell wall formation in fungi by competitive inhibition of chitin synthase (Gooday, 1995). Validamycin, which has an aminoglucoside structure, on the other hand, acts as a paramorphogen and inhibits infection cushion formation, an important step in the infection process, thus protecting the plant without necessarily killing the pathogen.

The first new fungicide, derived from a natural compound by fermentation, has been introduced by Dow, called fenpicoxamid. It has shown outstanding performance on *Septoria* spp. and has a broad spectrum of activity on rusts and other key cereal diseases. It is the first member of a new class of cereal fungicides called picolinamides and inhibits fungal respiration in the mitochondria at the Qi ubiquinone binding site in complex III, thus differing from all other cereal fungicides.

Other unusual fungicides include tricyclazole, a triazolobenzothiazole (from 1975), pyroquilone (1985) and carpropamide (1997), which are used in Asia to control rice blast. Of these, the mechanism of action of tricyclazole has seen the most attention. It inhibits the synthesis of fungal melanin in dark-pigmented ascomyctes. Pathogens such as *Magnaporthe grisea* and *Colletotrichum* spp. possess pigmented appressoria in which melanization is essential to cell wall rigidity and prevents bursting under osmotic stress. On treated plants, melanization is prevented and the infection process fails. Probenazole (1979), also used on rice against blast, was one of the first compounds identified to operate by increasing host resistance. Subsequently, research on systemic acquired resistance in plants and how it could be triggered chemically, led to the discovery of the benzothiadiazole 'plant activators' of which acibenzolar-S-methyl (1996) is an example. It is approved within the EU, but use in the UK is only until the end of 2018. Quinoxyfen is a quinoline marketed in 1997 to control powdery mildew on cereals and is also used off-label on soft fruit.

Mechanisms of Action

Studies on the mode of action are a key aspect of the modern development of fungicides. A general distinction can be made between compounds that affect multiple metabolic steps within target organisms (multi-site inhibitors) and those that affect only a single (usually enzymic) step or pathway (single-site inhibitors). As a broad generalization, the early fungicides to be introduced, up to around 1960, are multi-site inhibitors, although this was not fully appreciated at the time. Copper and mercury form complexes with sulphydryl, amino, hydroxyl and carboxyl groups of enzymes, so it is readily seen how a range of cellular functions are disrupted. The dithiocarbamates and phthalimides are also reactive with sulphydryl groups and disrupt the normal functioning of cellular glutathione.

The majority of chemicals introduced subsequent to the 1960s are single-site inhibitors. Some target very clearly defined steps; for example, the demethylation step in ergosterol biosynthesis; others are slightly less well defined, such as the inhibition of complex II in the mitochondrial electron transport chain. Reference is made in the section above to the mechanism of action of many of the major groups and a good summary is provided by Hewitt (1998). There is a helpful degree of correspondence between mechanism of action and fungicide chemical grouping, although there are some anomalies such as the unrelated aromatic hydrocarbon fungicides and the dicarboximides sharing cross-resistance and probable mode of action. Some modern fungicides target metabolism that is common to many organisms, in which case specificity might be limited. Others interact with metabolism that is unique to fungi, examples being chitin synthesis and the transformations leading to ergosterol synthesis. Utilizing such targets should provide a greater degree of selective toxicity and a larger safety net over effects on other organisms. There may also be less desirable consequences such as limiting the range of pathogens covered; for example, the absence of sterol synthesis by oomycetes explains their insensitivity to DMI fungicides.

Many modern fungicides are active at very low dose rates, which helps limit the problem of any residues in harvested product. The development of plant protection products that are not directly fungicidal but which work by stimulating plant resistance is a further approach to reducing the exposure of consumers to field-applied toxicants. Helpful though this trend may be, it does not guarantee acceptance of such products by regulatory bodies.

As discussed later, fungicide use has to be part of an integrated crop and pest management programme involving routine monitoring so that a fungicide is applied when and where the risk or presence of disease warrants treatment, before an infection is well established.

Disease Problems

Before fungicides were available, there were some major disasters as a result of not being able to control rampant diseases on certain crops. The Irish famine was due to late blight on potatoes, a disease of the foliage and tubers, causing rotting, which is most common in wet weather. The disease, caused by *Phytophthora infestans*, also attacks tomato plants.

Weather forecasts using Smith periods have helped to identify when conditions favour the pathogen, but recent studies reveal that certain dominant genotypes can infect potatoes at temperatures below 10°C and in high-humidity periods of less than 11 hours. Timing of applications is crucial, but good coverage is also important as spraying downwards over the crop, while covering new growth, may leave lower leaves still exposed to infection. Growing main-crop potatoes under European conditions is extremely difficult without fungicides because even late season minor outbreaks of blight can lead to infection of the tubers and rotting in store. Breeding for resistance to late blight has had mixed fortunes although some modern varieties with broad-based resistance do allow cropping with much-reduced fungicide inputs.

Coffee

The fungal disease coffee leaf rust, caused by *Hemileia vastatrix*, decimated the coffee industry in Sri Lanka in the 19th century and then spread to arabica coffee areas in southern India and other coffee growing areas in south-east Asia. Coffee leaf rust is of little significance at higher, cooler altitudes (>1700 m in equatorial areas), such as in east Africa and parts of South America, but the disease is now endemic in all major coffee-producing countries and requires control wherever arabica coffee is grown under warm, humid conditions. It spread to Brazil in 1970, then Mexico by 1981, and has resulted in significantly lower yields and crop quality in recent years in Colombia and central America. Climatic factors have favoured the disease, especially where there is a high density of coffee plants. Applying fungicides to coffee is difficult due to tree density and often steep topography, and usually relies on manually carried sprayers. Research had shown the superiority of applying copper oxychloride (30 droplets/cm^2) on the underside of leaves with an air-assisted, controlled droplet application knapsack sprayer (Motax) (Fig. 5.1) over conventional spraying when applied at an interval of 30 days between treatments (Waller *et al.*, 1994). Unfortunately, the importance of getting the fungicide within the crop canopy to minimize losses due to rainfall has not been recognized.

Cocoa

Problems continue for another tropical crop, cocoa, where black pod is caused by three species of *Phytophthora*, namely *P. palmivora, P. megakarya* and *P. capsici*. Global yield loss of 20–30% and tree deaths of 10% annually are estimated to be caused by *P. palmivora* alone (Fig. 5.2). In Africa, *P. megakarya* is the most important species as it is the most aggressive of the pod rot pathogens. In central and South America, *P. capsici* is widespread, causing significant losses in favourable environments.

Fig. 5.1. Spraying coffee with a Motax sprayer. (Photo courtesy of Micron, used with permission.)

Fig. 5.2. Spraying cocoa in Cameroon.

Another fungus known as 'frosty pod rot', caused by *Moniliophthora roreri*, occurs in all north-western countries in South America. This fungus has now spread all over the Latin American region, causing significant losses in production, including the abandonment of cocoa farms. Cocoa farmers in Bahia, Brazil, have now suffered a major drop in yields due to witches' broom disease, caused by *Moniliophthora perniciosa*, which has spread throughout all of South America, Panama and the Caribbean. Protectant sprays of copper-based fungicides, combined with the systemic fungicide metalaxyl under high-disease pressure, applied at three- or four-weekly intervals, are recommended, but where cocoa is exported to the EU, growers have to ensure that their cocoa beans do not exceed the maximum residue level (MRL). However, the greatest difficulty is for farmers with small areas of cocoa to apply fungicides effectively, particularly to pods in different parts of the crop canopy.

Soybean Rust

In 2001, soybean rust caused by *Phakopsora pachyrhiza,* long endemic in south-east Asia, was identified in Brazil. It has spread rapidly, assisted by favourable climatic conditions, and is now well established in the USA. This has resulted in an estimated annual loss from 2002 to 2012 of $3.8 billion. Generally, about two to three fungicide applications were made per season, initially with a triazole but subsequently with a triazole plus strobilurin mixture, as the pathogen became less sensitive to the triazole. More recently, crop protection companies have released fungicides with three active ingredients; thus Trivapro, from Syngenta, contains solatenol, strobilurin and a triazole. Ideally, there is a need for plant breeders to select varieties less susceptible to the disease, as using mixtures of fungicides can result in fewer modes of action that will be effective.

Cereals

Prior to the 1970s, the use of fungicides on cereals was unheard of, partly for reasons of economics, but also because the chemicals available were not good enough or available on a sufficient scale. However, trial application on a small scale had shown that some of the major foliar diseases could be controlled and they also revealed the scale of crop losses. From the early 1970s, under European conditions, it became practical to use chemical disease control. Applied as a seed dressing, carboxin was used to control loose smut, and ethirimol gave control of powdery mildew during early crop growth. Foliar applications of triadimefon (followed by other azoles and morpholines) were used to control leaf diseases, in particular yellow rust on wheat and brown rust on barley, and carbendazim (and later prochloraz) were used against eyespot disease. A useful summary of practice in the 1980s is given by Attwood (1985). Under European conditions of well-watered fertile soils with high inputs of fertilizer (and guaranteed prices), the use of fungicides augments other management practices such as crop rotation and the use of resistant varieties in giving good disease control, and is economically worthwhile. Although the list of fungicides changes over time, use on small-grain cereals (including rice) has become standard under high-yield potential cropping systems.

A recent evaluation in Ireland with varieties showing different septoria resistance treated with varying rates of an azole only or azole plus SDHI fungicide indicated that, when averaged over all sites and seasons, the income after fungicide costs was greater with the mixture for all varieties. It was considered that when growing new varieties with strong septoria resistance, a reduction in the fungicide programme may be possible.

Key diseases on wheat that are targeted are septoria (*Zymoseptoria tritici*), yellow rust (*Puccinia striiformis*), brown rust (*P. triticina*) and head blight (*Fusarium* and *Microdochium* spp.). The Agriculture and Horticulture Development Board organizes trials comparing individual fungicides to assess how each is effective against one disease, but the advice will largely be to apply a mixture of two to three fungicides for a limited period when disease is expected, to minimize selection for resistance. There is quite a problem for farmers with fungicides with different modes of action to know what combinations are needed. Against septoria, early treatment is important when growing winter wheat. Special nozzles, such as the Amistar nozzle, an air induction nozzle, was developed specifically to angle the spray 10° backwards. The aim was to get a better coverage and deposition on the stem as well as the leaves when applying usually 100 l/ha. Drift reduction was by using an air-induction rather than a standard nozzle type.

Rice

Rice has been subject to the disease rice blast (*Magnaporthe grisea*) at all phases of plant growth, but much has been achieved in several rice growing countries to avoid the use of fungicides by breeding varieties resistant to the disease. Resistance of a new rice cultivar to the disease usually breaks down within three or four years after the cultivar is released, so chemical control has been required in many areas. Various fungicides have been tried, but resistance of the pathogen has inevitably occurred. One unusual study reported that applying a spray of potassium silicate at 4 g/l reduced infection following artificial inoculation with a spore suspension five days after the silicate spray (Buck *et al.*, 2008). According to Yamaguchi (2004), two groups of non-fungicidal rice blast chemicals are currently on the market. As mentioned above, the melanin biosynthesis inhibitors (MBIs) tricyclazole, pyroquilon, carpropamid, diclocymet and fenoxanil have been used as well as probenazole, acibenzolar-S-methyl and tiadinil, which induce host resistance against the pathogen's attack. Among the fungicides evaluated, a mixture of tricyclazole with hexaconazole applied three times at weekly intervals starting at the booting stage gave effective control and the highest yield in a trial in Nepal (Magar *et al.*, 2015).

Bananas/plantains

Applying fungicides to bananas has presented special problems as the aim has been to protect the young leaves from infection with black leaf streak (BLSD), also known as black Sigatoka *Mycosphaerella fijiensis*, the name being associated with its detection in Fiji. Without control, leaf damage can be so severe that yields can be reduced by 35–50%. Uneven ripening also affects the suitability of a crop for export. Fungicides are best applied during the pre-necrotic stages of the disease. *Mycosphaerella musicola* (Sigatoka leaf spot, also known as yellow Sigatoka) occurred in Latin America and the Caribbean but has been largely displaced by black leaf streak in many banana production areas. In the 1930s, Bordeaux mixture was tried using ground equipment, but the youngest leaf is at the top of the plant, so getting a deposit on the undersurface as it unfurls is virtually impossible using ground equipment, as the large older leaves act as a barrier. New fungicides in the 1950s, plus the use of aircraft, enabled much lower volumes to be applied (Figs 5.3, 5.4, 5.5).

Fig. 5.3. Leaf showing Black Sigotoka. (Photo courtesy of John Clayton at Micron Sprayers, used with permission.)

Fig. 5.4. Rotary atomizer used on aircraft to spray bananas. (Photo courtesy of John Clayton at Micron Sprayers, used with permission.)

Fig. 5.5. Spray coverage on banana leaf. (Photo courtesy of John Clayton at Micron Sprayers, used with permission.)

Protecting new leaves led to frequent sprays, and adding a mineral oil with contact fungicides to enhance retention on the leaves led to phytotoxicity under hot, dry conditions. The main chemical classes of contact fungicides used in banana plantations have been benzene derivatives and carbamates. Chlorotalonil is a broad-spectrum contact fungicide, which has been used to control BLSD, while mancozeb has also been used. In the 1960s, benzimidazole fungicides were introduced as their systemic activity enhanced control, but using these too frequently led to the pathogen populations becoming less sensitive or resistant to them. The FRAC banana working group focusing on the control of BLSD has recommended eight main classes of fungicides to control fungal diseases, including strobulirins, sterol demethylation inhibitors and others. They advocated using mixtures and avoiding more than three sprays per season using a fungicide with any one mode of action, and separating application of some modes of action by a free period of three months. To reduce selection pressure for resistance, they also recommended that the total number of applications per year of a mode of action should not exceed eight and should not represent more than 50% of the total number of sprays.

Fruit

Apple trees were treated with Bordeaux mixture to control scab in the 1880s, although the copper can cause russeting on the fruit. Chloride of iron was also applied prior to bud formation. With IPM there is now much emphasis on removing any leaf litter during the autumn to avoid transfer of spores to the following year, as ascospores of *Venturia inaequalis* from leaf litter are easily spread to the young foliage in the spring. Monitoring of the orchard and using a disease forecasting system is advocated to minimize fungicide use in an integrated approach using cultural control to harvest scab-free fruit.

Fig. 5.6. Air-assisted sprayer treating vineyard. (Photo courtesy of Technoma, used with permission.)

Grapes

The need to protect vineyards from diseases was clearly established in the 19th century with the development of Bordeaux mixture to control downy mildew (*Plasmonara viticola*) and powdery mildew (*Erysiphe necator*), the most important grapevine diseases. Together with control of grey mould (*Botrytis cinerea*), these three diseases, which affect the leaves and can spread to the fruit, account for the largest number of spray applications on grapevines.

Sulfur is still used to control powdery mildew, but growers need to be careful as it should only be used on varieties tolerant to applications; but it has little effect on other diseases. Mancozeb has been used on grapes, but as with other crops, it is better to rotate its use with other fungicides to avoid selecting resistance in the fungus. Strobilurin fungicides fit in such a rotation. Another disease is black rot (*Guignardia bidwelli*), which attacks plants during hot and humid conditions. The preferred fungicide is either captan or myclobutanil. Mancozeb has been used but the pre-harvest spray interval is 66 days, which limits its use.

The number of applications has varied depending on climatic conditions and the variety and severity of the disease, but fungicides can be applied up to 25–30 times in worst-case situations (Pertot *et al.*, 2017). A general rule is that it is unwise to apply more than two sequential sprays of any material that is at risk for resistance development, before alternating to a fungicide with a different mode of action. Resistance occurs with the more specific fungicide or class of fungicide used in a vineyard, so when these are used it is essential to limit the number of applications that can be made per season (Fig. 5.6).

Rubber

When they decided to grow rubber trees in Malaysia they obtained the initial plants from Brazil via Kew Gardens, in quarantine, as a precaution, to avoid taking the disease South American leaf blight (*Microcyclus ulei*) on the plants. In Brazil the disease can spread rapidly via spores, which results in the fungus causing premature leaf fall leading to dieback of trees and economic losses (Fig. 5.7). Some plants are resistant but these usually have a low output of latex. Various crosses have been used to try to improve yields by plant selection. Rubber growing in other countries has been successful in the absence of the disease. In Brazil, thermal fogging (Fig. 5.9) was carried out on rubber estates as using a mist failed to treat large areas effectively. As the youngest leaves are the most susceptible, aerial treatment was recommended (Fig. 5.8).

Fig. 5.7. Area of Brazil with South American blight defoliation of rubber trees.

Fig. 5.8. Young rubber tree leaves most prone to the disease.

Fig. 5.9. Thermal fog applying fungicide on a rubber estate in Brazil. (Photo courtesy of Tifa, used with permission.)

References

Attwood, P. (ed.) (1985) *Crop Protection Handbook – Cereals*. British Crop Protection Council, London.

Buck, G.B., Korndörfer, G.H., Nolla, A. and Coelho, L. (2008) Potassium silicate as foliar spray and rice blast control. *Journal of Plant Nutrition* 31, 231–237.

Gooday, G.W. (1995) Cell walls. In: Gow, N.A.R. and Gadd, G.M. (eds) *The Growing Fungus*. Chapman & Hall, London, pp. 43–62.

Hewitt, H.G. (1998) *Fungicides in Crop Protection*. CAB International, Wallingford, UK.

Magar, P.B., Acharya, B. and Pandey, B. (2015) Use of chemical fungicides for the management of rice blast (*Pyricularia oryzae*) disease at Jyotinagar, Chitwan, Nepal. *International Journal of Applied Sciences and Biotechnology* 3, 474–478.

Pertot, I., Caffi, T., Rossi, V., Mugnai, L., Hoffmann, C. *et al.* (2017) A critical review of plant protection tools for reducing pesticide use on grapevine and new perspectives for the implementation of IPM in viticulture. *Crop Protection* 97, 70–84.

Waller, J.M., Lequizamon, J., Gill, L.F., Aston, R.A., Cookman, G.F., Sharp, D.G., Ford, M.G. and Salt, D.W. (1994) Laboratory and field development of a CDA spraying system for control of coffee leaf rust (*Hemileia vastatrix*): an overview. *Comparing Glasshouse and Field Performance II*. BCPC Monograph 59, 261–266.

Yamaguchi, I. (2004) Overview on the chemical control of rice blast disease. In: Kawasaki, S. (ed.) *Rice Blast: Interaction with Rice and Control*. Proceedings of the 3rd International Rice Blast Conference, 1–13. Springer Nature.

6 Other Pesticides

Rodenticides

Apart from the damage caused by rats to food in storage on farms and in markets, damage to crops in the field can also be extensive in many parts of the world. Red squill was extensively used to control rats and mice, but had little effect on other mammals as it is a strong emetic and was regurgitated before it could do any harm. Strychnine, which is generally toxic to warm-blooded animals, was also used prior to the development of warfarin. Apart from rats, strychnine was also used to control moles, but this is no longer allowed.

Warfarin, a coumarin, interferes with the action of vitamin K in the body, reducing the ability to coagulate blood by reducing the prothrombin content of the blood, thus causing internal bleeding and a fatal haemorrhage if the body was cut. It was shown that a daily dose of 2 mg was fatal to rats, whereas dogs survived daily doses of 50 mg.

Warfarin is prescribed for humans to prevent thrombosis, the formation of life-threatening clots in arteries or veins, but since about 1944 it has become a major rodenticide. As with other pesticides, rats have become resistant to it, and in the 1950s several other more toxic rodenticides were marketed. These included fumarin, diphacinone and chlorophacinone. In the 1960s, pindone was a new anticoagulant, which did not lead to bait shyness and so was effective if the rats fed over several days and accumulated a lethal dose. Coumatetralyl was also marketed in 1962.

With resistance to anticoagulants, it was fortunate that a second generation of more active rodenticides were marketed and used at much lower doses. These included difenacoum, brodifacoum and bromadiolone, followed later by difethialone and flocoumafen. Although death can occur after a single feed of the rodenticide, it may not occur for 3–4 days. Although this reduces any impact of bait shyness, there has been increasing concern about the effect on wildlife that consumes dead or weakened rodents. Environmentalists prefer people to use low-strength, persistent rodenticides, generally the first anticoagulants, on the assumption that these pose less risk to wildlife. The death of an animal can be due to deliberate illegal use of a rodenticide, incorrect procedure in using a rodenticide or by accident, if not following all the label instructions. Strict protocols have been published to minimize the risk of non-target animals being affected. In the UK, there is a Campaign for Responsible Rodenticide Use (CRRU), which promotes training of those using rodenticides, best practice and responsible rodent control, thereby protecting wildlife from rodenticide exposure.

In the USA, the EPA imposed restrictions on the sale and distribution of rodenticides, which must be provided in tamper-proof packaging to reduce the risks when using them.

Rodenticides are usually used in baits and a survey of 956 baits sampled in one area of Italy revealed not only 11 different rodenticides being used, but that 9.3% of the baits had more than one toxic compound. These were mostly highly toxic insecticides such as methamidophos, carbofuran and endosulfan, which are banned in the EU (Chiari *et al.*, 2017).

Although there has been continued controversy about the use of sodium fluoroacetate (Compound 1080), it has been used throughout Australia since the early 1960s as it is regarded as a very useful pesticide for the control of invasive animals. Indigenous animals are said to tolerate the poison, which occurs in Australia naturally in plants. It is regarded in Australia as a species-specific pesticide currently available for invasive animal control, including wild dogs, feral pigs, foxes, feral cats and rabbits, to protect agricultural production and native flora and fauna from the impacts of invasive animals. The use of 1080 in some conservation areas allows the continued survival of rare and threatened wildlife and assists in the reintroduction of species into areas where they have previously been extinct locally. While in many parts of Australia it has been applied aerially as a bait, treatment of smaller areas in New Zealand using aircraft has raised questions about the safety of people in such areas.

A risk analysis in Australia indicated that if 48 mg of Compound 1080 were applied/ha (by using a wild dog bait containing 6 mg/kg applied at 8 kg/ha) and all the rodenticide was washed by heavy rain, an individual person would need to drink over 16,000 litres of contaminated water to get a lethal dose. Similarly, if a 60-kg feral pig ingested 3 kg of a bait, thus consuming 1152 mg/kg of bait, and assuming half the ingested poison was evenly distributed in the pig's carcass, the hunter who shot the pig would need to eat 36.1 kg of meat in one sitting to be at risk (Anon, 2017).

Molluscicides

Slugs (*Deroceras reticulatum, Arion fasciatus, Deroceras leave*) and snails can cause considerable loss of crops immediately after sowing. This can be by hollowing out seeds of crops such as cereals before germination (i.e. before the plants are above the surface) or by feeding on young/small plants immediately after germination in crops such as oilseed rape (Fig. 6.1). To combat this, farmers have used molluscicides to protect their crops. Slug activity is difficult to detect as a large proportion of the slug population is often located within the upper layers of soil and is not visible. Traps can be deployed to attract the slugs and determine whether there is a risk of crop damage, which will also be influenced by weather forecasts for rain, but the traps assess the number of slugs that are active on the surface so can result in less than accurate assessments of slug populations and economic damage. With traps situated in areas to be sown with crops, such as wheat and oilseed rape, four slugs per trap have indicated a possible risk of slug damage. Studies have shown that there is less damage if the seed is sown at 4 cm below the surface rather than at a shallower depth (Glen *et al.*, 1990), but where there is a high risk, slug pellets need to be applied as soon as possible after drilling as each slug, especially *Deroceras reticulatum*, can kill up to 50 seeds in the first week after sowing, with the smaller slugs killing more seeds than larger slugs (Fig. 6.2). Further trapping is advised until Growth Stage 21 (GS21) as the young seedlings can also suffer damage.

In 1936, in southern France, a solid fuel sold as meta-tablets was used by some campers and some of the tablets were left on the ground. It was then noticed that there were dead slugs in the same area, thus the molluscicidal activity of metaldehyde was noticed. This cyclo-octane, approved in the EU, kills the slugs by contact and stomach poisoning that stimulates mucus/slime production resulting in desiccation of the slug.

The problem for farmers is which molluscicide can be used, as it is important to avoid polluting water. Slug control has, historically, relied on a limited number of active ingredients with methiocarb and metaldehyde dominating the market. Methiocarb was withdrawn from the market in 2015 and metaldehyde now accounts for a large proportion (84%) of slug treatment in the UK, treating nearly a third of the wheat area and half of the rapeseed area (Garthwaite *et al.*, 2015). Metaldehyde is subject to restrictions and best practice guidelines aimed at protecting water courses. The level of metaldehyde must be below 0.1 µg/l in drinking water, so a buffer zone of at least 5 m was needed between areas treated

Fig. 6.1. Slug.

Fig. 6.2. Applicator for slug pellets.

with this active ingredient and a water course or any field boundary, but in 2017 this was increased to 10 m. Treatments must also be avoided when heavy rain is forecast. The aim is to avoid metaldehyde being detected in untreated water at levels above the drinking water standard, which has occurred in the past.

Currently no more than 250 g of metaldehyde can be applied/ha per application, with some recommending a lower dose of 160 g a.i./ha and no more than 700 g a.i./ha/year. Pellets with a bitter taste are required to avoid them being consumed by non-target animals. Where land is sloping, farmers are advised to use the alternative product ferric phosphate.

The parasitic nematode *Phasmarhabditis hermaphrodita* has been effective as an alternative to chemical molluscicides, but is used mostly on small areas in gardens or high-value horticultural crops due to cost. The nematodes require moist soil and an application rate of 300,000 infective juveniles/m^2 applied as a drench or spray.

Snails (*Bulinus* and *Biomphalaria* spp.) are also the vector of the disease schistosomiasis, referred to in some countries as bilharzia, caused by parasitic flatworms that infect the intestines. People bathing in rivers or other wet areas where the snails occur will get infected by the worms through the skin. The WHO requested development of a knapsack sprayer to deliver a set volume of liquid for each pump stroke to project the molluscicide at intervals along a waterway (Anon, 1990). The molluscicide niclosamide was assessed in various countries but, as reported in St Lucia, although the effects of consistent multi-year treatments were clearly beneficial, total elimination could not be achieved (Sturrock *et al.*, 1974). The cost of labour to treat areas with niclosamide was high, so when the drug praziquantel became available, attention was given to mass drug administration. Praziquantel given as a tablet annually has been recommended, but now with concerns about resistance to the drug, more attention is being given to control of the vector (King and Bertsch, 2015). In 2015, an estimated 252 million people worldwide were affected by schistosomiasis. In tropical countries, schistosomiasis is considered, after malaria, to have the greatest economic impact among parasitic diseases.

Snails have been causing major problems in rice. The golden rice apple snail (*Pomacea canaliculata* and *P. maculata*) is highly invasive and causes damage to young rice plants. It was introduced from South America to be used in restaurants, but spread rapidly along irrigation lines in south-east Asia (Joshi *et al.*, 2017). A similar snail problem is in Spain, where *Pomacea insularum* was reported in 2009, with up to half of crops affected by as many as 12 snails/m^2 in the worst areas. Damage is caused by the snails eating the roots as well as the young leaves. Cultural control with better water management was recommended, with continuous hand picking, trapping and destruction of egg masses, but where necessary, a molluscicide can be used to protect young seedlings. In Spain a natural extract from camellia seeds, saponin, has been authorized in efforts to stop the spread of the snails.

Fumigants

Fumes of sulfur were used as a fumigant as early as the 12th century BC, but in 1869 carbon disulfide was used against the grape phylloxera, a landmark in the history of applied entomology. It was injected into the soil to control the insects infesting the roots of the grapevine. Carbon disulfide was then widely used as a soil or space fumigant, but could explode, so it was formulated with carbon tetrachloride and now with a mixture of non-flammable ingredients to fumigate grain. It is still used, as the gas penetrates well, especially in tropical countries where the high temperatures favour volatilization. In the early days of fumigation, a number of other chemicals were tried, including trichoroacetonitrile, used as a louse fumigant in Germany; ethylene dichloride; methanesulphonyl fluoride, soon considered too hazardous to use; and acrylonitrile, which was used with carbon tetrachloride to kill insects in flour mills and bed bugs.

Hydrogen cyanide was one of the first fumigants to be used extensively under modern conditions. Its use for treating trees under tents against scale insects was developed in California in 1886 (Woglum, 1949). The use of HCN has been declining in recent years, but it is still important in certain fields of application.

Fumigation can be done in a variety of situations, including specially built fumigation or sterilization chambers, or outside in temporarily created containment situations, for example in stacks of maize covered and sealed with tarpaulins (Figs 6.3, 6.4). During World War II, fumigants came into prominence in protecting food in storage and transit. A considerable amount of research took place at the Imperial College Field Station in Slough, which became the Pest Infestation Laboratory after the war ended. Page and Lubatti (1963) reviewed the studies on fumigation after the field station had been moved to Silwood Park. In some situations, in sealed environments, it has been possible to control the atmosphere within a sealed silo or store by adding nitrogen and/or carbon dioxide to prevent survival of pests instead of using a fumigant.

Fig. 6.3. Large modern silo in which grain can be protected by fumigation.

Fig. 6.4. Preparing to add fumigant under the plastic sheeting. (Photo courtesy of Rentokil.)

Consignments moved around the world by the trade in plants and plant products often are infested with insects, so fumigation of these is warranted to meet plant health/quarantine regulations and to prevent pests becoming established in a new geographical area, since this can be achieved without undue delays in the movement of sometimes highly perishable goods. In some cases, produce for export has also been fumigated onboard ships.

Fumigation of soil, especially seed beds, covered by an impermeable plastic sheet, has been an established technique for certain crops, notably tobacco and strawberry production (Figs 6.5, 6.6). Improved plastic sheeting reduced the loss of the fumigant into the air during treatment.

Ethylene dibromide

Ethylene dibromide (EDB) was used extensively as a soil and post-harvest fumigant for crops, and as a quarantine fumigant for citrus and tropical fruits and vegetables. In 1983, the EPA suspended the use of EDB as a fumigant when low-level residues were found in groundwater and some grains.

D-D mixture

In 1943, in field experiments with a mixture of 1,3-dichloropropene and 1,2-dichloropropane in pineapple fields in Hawaii, the material was reported to be less expensive and easier to handle than chloropicrin, while producing comparable results controlling nematodes and insects. It was subsequently used by injecting controlled dosages into soil as the fumigant action was effective against plant parasitic nematodes attacking a broad range of crops. It was discontinued in the 1980s.

1,3-dichloropropene

This component of D-D was used as Telone, produced by Dow Chemical until 1990 when it was detected in ambient air in California. However, it was later reintroduced with strict control measures in 1995.

DBCP

DBCP (1,2-dibromo-3-chloropropane) was developed by McBeth and Bergeson (1955) presumably as it was not highly toxic to many crop plants and could be applied at planting or after planting. In 1964, the US government approved DBCP for commercial use and the companies proceeded to market the pesticide, but apparently did not divulge its full 'extremely hazardous' toxicity or recommend protective clothing. Commercialized as Nemagon, it was extensively used for control of citrus-root nematodes (*Tylenchulus semipenetrans*) in the USA and for control of other nematodes on living trees and grapevines. It was also used on soybeans. At its peak, 426,000 pounds were used in California in 1977, but its use was stopped when it was discovered that it caused infertility and sterility in workers who formulated the product. It was also estimated that as many as 500,000 Californians had DBCP in their drinking water supply. In central America it was also widely used as a nematicide on bananas until 1977 when workers and their union at a formulating plant in Occidental, California, identified the first human sterility cases linked to DBCP. The product was subsequently banned (see also Chapter 9).

Metam

Sodium N-methyl dithiocarbamate dehydrate, sold as Vapam, was introduced by the Stauffer Chemical Company in 1955. It decomposed in the soil to form a penetrating gas, and thus acted as a fumigant, effective for control of nematodes, weed seeds and soil fungi.

Methyl bromide

Methyl bromide (bromomethane), developed around 1940 as a fumigant, has been the most widely used fumigant for quarantine purposes, as the gas rapidly penetrates the space being fumigated and displays high toxicity to a broad spectrum of insects and pests. As it was considerably safer and more effective than using other methods of sterilizing soil, it was also widely used as a soil sterilant. Chloropicrin, developed as a fumigant in 1936, has a distinctive odour and was added to the methyl bromide to enable any escaping gas to be readily detected. On tobacco farms, the seed bed was covered by a plastic sheet with the edges buried to prevent gas escaping (Figs 6.5 and 6.6). A canister of gas was placed under the sheet so that a hole could be punched in the can when it was covered by the sheet, but this was changed. The use of methyl bromide has now been curtailed by the Montreal Protocol. However, with over 7 million pounds of bromomethane applied in California in 2004 for certain crops, notably tomato and strawberry, and treatment of solid wood packaging/ packaged goods, there have been efforts by the USA to obtain critical use exemptions. Because of the harmful impact of methyl bromide on the ozone layer, it was due to be phased out completely in 2017. The changing pattern of fumigant use in California shows increase in some alternative products as use of methyl bromide has declined (Fig. 6.7).

Phosphine

Use of phosphine has increased following greater restrictions on the use of methyl bromide. Phosphine is emitted from tablets of aluminium or

Fig. 6.5. Preparing to fumigate a tobacco seed bed with methyl bromide, Zimbabwe. (Photo courtesy of Kutsaga Research, used with permission.)

Fig. 6.6. Canister of methyl bromide ready for gas to go along tube and be released under the plastic sheet. (Photo courtesy of Kutsaga Research, used with permission.)

magnesium phosphide when exposed to air, so highly dangerous, unless strict regulations are followed to protect those who are fumigating grain in a silo to prevent loss of the gas to the atmosphere. The phosphine gas must be retained within the silo and circulated evenly through the bulk of grain for several days, as it may take up to three days to kill some resistant pests,

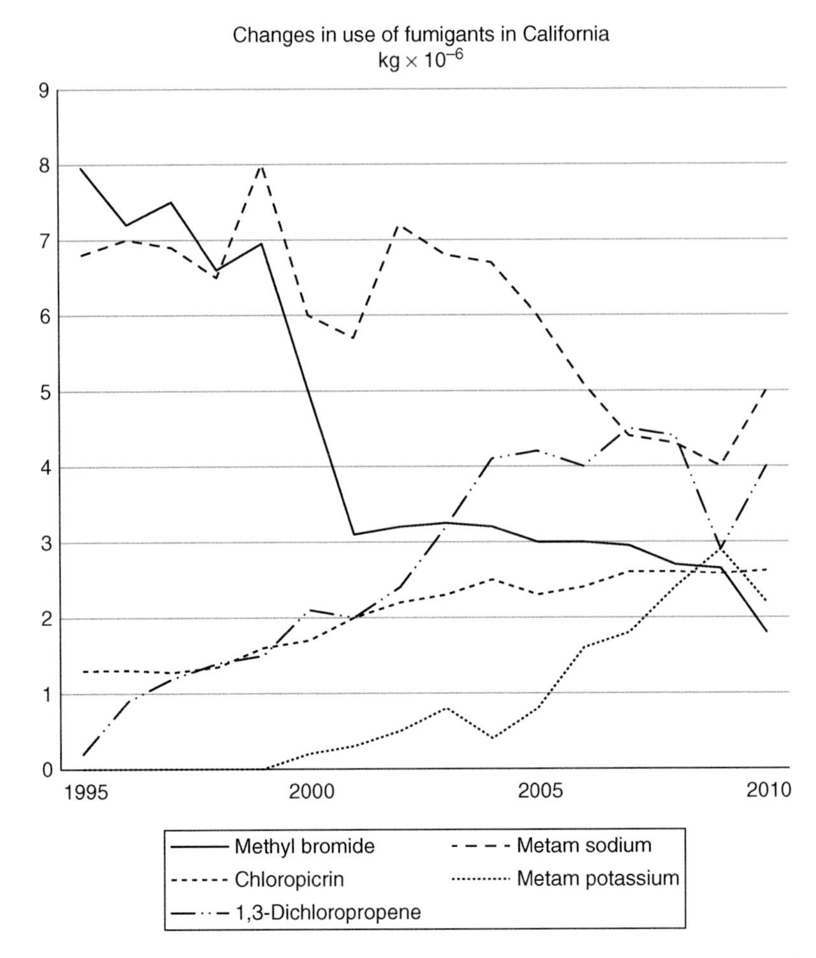

Fig. 6.7. Changes in use of fumigants in California. (Redrawn from Epstein and Zhang, 2014.)

and longer at low temperatures. There have been tragic accidents when tablets have been incorrectly used in tropical countries. To avoid using the tablets, specialized automatic systems have been developed to use phosphine in a cylinder and release the gas at different positions within the silo. Strict cleaning of stores and use of residual insecticide sprays on walls to prevent infestation during storage are essential to minimize any need for fumigation of the grain. As commodities are now often shipped in transport containers, specialized pest control companies can fumigate the containers.

Sulfuryl Fluoride

The Dow Chemical Company developed sulfuryl fluoride as a structural fumigant insecticide to control termites affecting dry wood, particularly in warmer climates. It can be used also to control rodents, bark beetles and bedbugs. It is now one of the replacements for methyl bromide and an alternative to phosphine, which is acutely toxic.

Other Nematicides

Fenamiphos (Nemacur) was introduced in the late 1960s as the systemic nematicide with contact activity. Fensulfothion (Basanit) was developed at the same time, but is no longer produced. Carbofuran (Furadan), ethoprop (Mocap), aldicarb (Temik) and oxamyl (Vydate) have all been registered to control nematodes, but only oxamyl is now registered in the UK. Tioxazafen was introduced in 2013 as a new broad spectrum nematicide and is approved as a seed treatment on cotton, maize and soybeans under a trade name NemaStrike.

Avicides

The use of pesticides generally in agriculture is often considered to be the cause of any decline in bird populations. The use of herbicides to control weeds reduces the seed availability for some birds, while control of insects reduces the number of larvae that birds can feed to their young. Rachel Carson was particularly concerned about the persistence of DDT in the environment and in food chains. The impact on birds was particularly severe as it resulted in thinner egg shells, so the egg could be damaged before the young chick had developed (see Chapter 10, 'Stockholm Convention').

In contrast, birds that damage crops in sub-Saharan Africa are regarded as pests. The major cause of damage is by the red-billed weaver bird, *Quelea quelea*, which is inherently nomadic following rain fronts, which enable the species to invade areas where it has not previously been present (Ward, 1971). They move away from their dry season habitat towards areas where rain had started several weeks earlier. The direction taken by the migrants, the distance they must fly and the timing of the movement are dependent upon the timing of the rains and the way the rain front moves. The birds aim to be able to find crops such as sorghum just before the crop is ready to harvest. Other small grain crops are also attacked, but crops such as maize are avoided as the seeds on the cob are protected by the outer sheath around the cob. Swarms of the birds will have thousands of individuals migrating at the same time. Their feeding can rapidly cause major yield loss for the farmer, so efforts have been made to control them throughout east Africa. Small areas with the weaver birds have been tackled by using nets, but the application of an organophosphate spray has been the preferred method of control. Small areas can be sprayed with a knapsack mistblower, such as the AU8000 with a rotary nozzle, but large populations are treated using an aircraft. This can be directed at a swarm of birds, but is often applied as the birds are flying into a roosting site at dusk. Fenthion has been used, although use of it is now very restricted, and cyanophos has been suggested as an alternative (Cheke, 2016), although further research on the environmental impacts of cyanophos is recommended.

References

Anon (1990) *Equipment for Vector Control.* World Health Organization, Geneva, Switzerland.

Anon (2017) *Sodium Fluoroacetate (1080). Fact Sheet.* The State of Queensland, Department of Agriculture and Fisheries, Australia.

Cheke, R.A. (2016) Alternatives to Fenthion for Quelea Bird Control. Report of the UNEP-FAO-RC-Workshop, Sudan.

Chiari, M., Cortinovis, C., Vitale, N., Zanoni, M., Faggionato, E., Biancardi, A. and Caloni, F. (2017) Pesticide incidence in poisoned baits: a 10-year report. *Science of the Total Environment* 601–602, 285–292.

Epstein, L. and Zhang, M. (2014) The impact of integrated pest management programs on integrated pest management in California, USA. In: Pestin, R. and Pimental, D. (eds) *Integrated Pest Management.* Springer, Dordrecht, The Netherlands.

Garthwaite, D., Barker, L., Laybourn, R., Huntly, A., Parrish, G.P., Hudson, S. and Thygesen, H. (2015) *Pesticide Usage Survey Report 263: Arable Crops in the UK 2014.* FERA, York, UK.

Glen, D.M., Milsom, N.F. and Wiltshire, C.W. (1990) Effect of seed depth on slug damage to winter wheat. *Annals of Applied Biology* 117, 693–701.

Joshi, R.C., Cowrie, R.H. and Sebastian, L.S. (eds) (2017) *Global Advances in the Ecology and Management of Invasive Apple Snails.* Philippine Rice Research Institute, Munoz, Philippines.

King, C.H. and Bertsch, D. (2015) Historical perspective: snail control to prevent schistosomiasis. *PLOS Neglected Tropical Diseases* 9(4), e.0003657.

McBeth, C.W. and Bergeson, G.B. (1955) 1,2-dibromo-3-chloropropane – a new nematocide. *Plant Disease Reporter* 39, 223–225.

Page, A.B.P. and Lubatti, O.F. (1963) Fumigation of insects. *Annual Review of Entomology* 8, 239–264.

Sturrock, R.F., Barnish, G. and Upatham, E.S. (1974) Snail findings from an experimental mollusciciding programme to control *Schistosoma mansoni* transmission in St Lucia. *International Journal of Parasitology* 4, 231–240.

Ward, P. (1971) The migration pattern of *Quelea quelea* in Africa. *Ibis* 113, 275–297.

Woglum, R.S. (1949) History of fumigation in California. *California Citrouraph* 35, 46–72.

7 Resistance to Pesticides

The occurrence of resistance in an insect pest was first reported in the literature in 1914, where treatments of lime sulfur had been carried out each year for 25 years in Washington State to control San José scale, *Quadraspidiotus pernicious*, which had entered the USA about 50 years earlier. The scale insect was now more resistant to the lime sulfur (Melander, 1914; Forgash, 1984). Melander predicted that entire populations would not become resistant as long as some non-resistant insects survived, because their non-resistant genes would be passed on to future generations. However, a pure resistant line might result after repeated sprayings, if only the resistant individuals survived to reproduce. He found that after 11 years, 74% of the scales survived, despite using a higher dose of lime sulfur.

Babers (1949) reported that soon after Melander's observations, the California red scale, *Aonidiella aurantii*, became more difficult to control with hydrocyanic acid fumigation and that the dosage required to control resistant strains was so high that it was unsafe for the tree, except in the most favourable conditions. Ripper (1956) noted that studies in 1929 using lime sulfur on citrus trees increased the number of red scales compared with untreated trees, but when Debach and Bartlett (1951) sprayed DDT, a dosage that was insufficient to control the red scale killed the parasites and predators, so the subsequent increase in the population of scales was due to severe disruption of the balance between the California red scale and its predator *Aphytis chrysomphali* rather than due to selection of a resistant strain.

The selection of resistance in an insect should have been foreseen, as during the Industrial Revolution, with coal-burning factories in England, the melanic form of the peppered moth (*Biston betularia*), which was rare in the early 1800s, was 98% of the population in woodlands near Manchester by 1895, due to the soot deposited on the trees. This was because birds and other predators could easily detect the lighter form of the moth on the blackened trees. Much later, after the Clean Air Act was passed, the selection pressure changed allowing the light form to survive. This exhibited Darwin's theory of natural selection, which underlies how the resistant insects can survive and flourish during a period of continuous exposure to an insecticide.

Between 1900 and 1950 there were few cases of resistance to insecticides being recorded, but as soon as DDT was used, resistance was detected in house flies as soon as 1947, and subsequently in many different insect pests. DDT had been used extensively in agricultural crops and in public health, notably for spraying the inside walls of houses to control the mosquitoes that transmitted malaria. DDT was particularly good for indoor residual spraying as the deposits remained effective for months. The global programme was curtailed when resistance to DDT was detected in the malaria vector *Anopheles* spp. in 1969. The resistance problem returned when further efforts to control malaria were promoted, and this is discussed later in the chapter.

Since spraying of plants with pesticides was adopted by farmers, inevitably there has been selection of the survivors whenever a pest population has been exposed to a pesticide for a prolonged period. This could be due to the survivors being able to metabolize the pesticide or, more commonly, having a mutation that makes them less sensitive to the pesticide (altered target site). Lack of control can also be due to avoiding the pesticide by a behavioural action or by being repelled by the spray deposit, or simply the effect of a poor application that did not deposit the spray where the pest was located. The speed of selection of a 'resistant' population would depend on the selection pressure and whether a localized effect is diluted by immigration from untreated areas. Thus frequent applications of a pesticide, or persistence of spray deposits with a specific mode of action and/or belonging to the same chemical group, over a large area against a pest, disease or weed, will inevitably result in selection of a resistant population. It will also occur rapidly if a pest population is confined, for example inside a glasshouse. This will be quicker with insects with a short generation period compared with a weed, where dormant seeds may germinate, be still susceptible and dilute the selection pressure. Increasing the dose, if the pest is not effectively controlled, will in all probability result in the pest becoming resistant quicker, and will at the same time cause much more harm to predators and other non-target organisms. The problem of resistance is now of increased importance with genetically engineered crops, with either tolerance to a specific herbicide or increased toxicity to insects that attack the crop.

Resistance to insecticides is frequently due to an enzyme that can break down the chemical; this is referred to as metabolic resistance. Target site resistance prevents the pesticide from being active within the pest insect. Multiple resistance is when insects have a means of detoxifying more than one type of insecticide. Cross-resistance is if a mutation allows an insect to be resistant to one insecticide and it is also resistant to another compound or compounds with the same mode of action (target site resistance) and/or similar chemistry (metabolic resistance). The level of resistance can be monitored by laboratory bioassay tests, and enzyme levels can also be measured.

Resistance Management

Several practical approaches have been recommended to counter resistance:

- developing a schedule that uses pesticides with a different chemistry and mode of action; this may involve limiting specific pesticides to only a limited number of applications per year, or, as in the acaricide rotation scheme discussed below, limiting use in one area for a period and then not using it for a longer period;
- alternating use of pesticides with different modes of action (but continued use of the pattern could lead to resistance to both modes of action);
- using a mixture of pesticides with different modes of action with similar persistence;
- having untreated areas ('refugia') to allow susceptible pests to survive and dilute the 'resistant' population.

The idea of a national or even regional schedule to use pesticides according to their mode of action has hardly been considered. New or least expensive chemicals are applied until resistance is clearly obvious. Perhaps this is illustrated by the continued use of glyphosate-tolerant crops of maize, soybeans and cotton over vast areas, even when there were obvious signs that weeds were no longer being controlled. A similar situation has occurred with insecticides, with a progressive use of new molecules as they became commercially available. Fig. 7.1 shows an example of this with sprays against the aphid *Myzus persicae* on cereals in the UK.

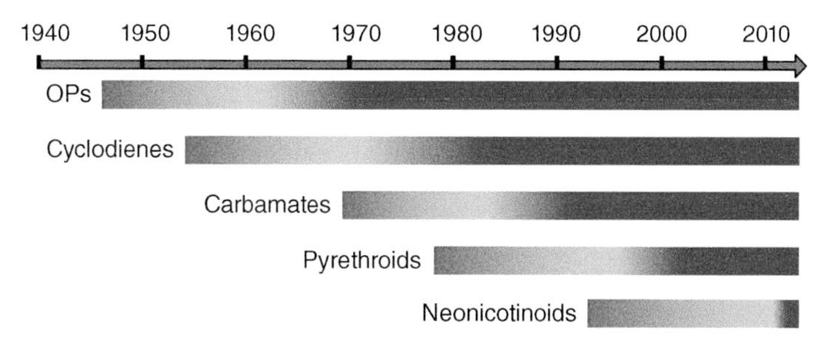

Fig. 7.1. Sequence of different insecticides used to control cereal aphids in UK. Green bars show when good control was achieved while red indicates when resistance adversely affected control. (Courtesy of Rothamsted Research. Published by Elsevier under a Creative Commons Licence.)

The following examples illustrate situations where pragmatic decisions were made to maintain control of the pest and mitigate against selection for resistance.

The Acaricide Rotation Scheme

In the 1960s, dimethoate had been recommended to control aphids and other sucking pests plus red spider mites on cotton in Southern Rhodesia (now Zimbabwe). By 1966, some farmers had reported that they were not controlling the red spider mites, but this seemed odd as the number of sprays of dimethoate on cotton was only two to three in a season; but it was then realized that the farmers who had a problem were using dimethoate throughout the year on a range of irrigated vegetable crops. Duncombe (1973) introduced an acaricide rotation scheme in which the country was divided into three zones and acaricides with different modes of action were allocated to each zone (Fig. 7.2). After two years the unused chemical was moved to the next zone, so the cycle was repeated after six years. This operated very effectively from the late 1960s until at least 2000, and spider mites were again susceptible to dimethoate.

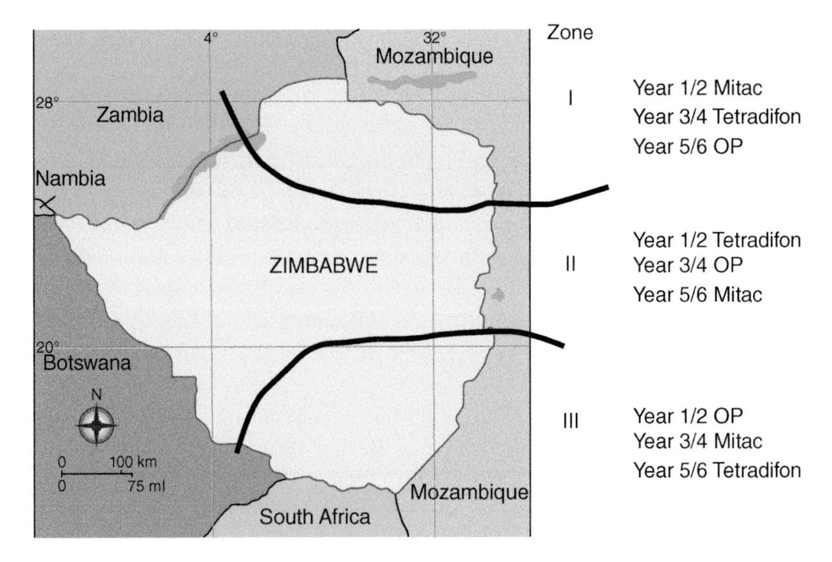

Fig. 7.2. Map of Zimbabwe showing the zones in which acaricides were rotated.

Restricted Use of Pyrethroids in Australia

In Australia, cotton farmers started to apply pyrethroids instead of DDT to control cotton bollworms, but soon there were reports of resistance in the 1980s. The recommendation was to limit the use of pyrethroid insecticides throughout Australia on any crop for a limited period on the assumption that the bollworms without exposure to selection for resistance to pyrethroids during the remainder of the year would be effectively controlled the following year. The situation then changed with the introduction of genetically modified, transgenic cotton varieties (GM cotton) that expressed an insecticidal toxin from the soil bacterium *Bacillus thuringiensis* (Bt) in order to kill first instar larvae as soon as they started to eat their way into a bud or boll. The problem is now more one of preventing bollworms developing resistance to Bt toxins, necessitating the use of preemptive resistance management strategies, including the use of 'refugia' and Bt cotton constructs expressing more than one toxin.

Alternating Pesticides or Use of Mixtures Approach

More often, the approach to reducing selection of resistant pests has been to recommend the application of a particular pesticide for only one or two applications per season and alternating with another pesticide having a different mode of action, or simply applying a mixture of both chemicals. The mixture is still likely to enable selection of resistance to both modes of action, although it may take much longer to see an impact on the level of control obtained. When carbaryl and DDT were recommended on cotton for different bollworms, one company immediately offered farmers a mixture, but its use was resisted and scouting of the crops promoted to ensure only one of the insecticides was applied in relation to the pest population present. The alternation of modes of application is likely to be better, but much depends on the duration of exposure to each mode of action within each generation. The agrochemical companies now put information on product labels, generally recommending the maximum number of applications in a single season to assist farmers in choosing a product with a different mode of action.

With the introduction of genetically engineered cotton, it was recommended that farmers had an 'untreated' crop or 'refuge' area to provide a source of susceptible pests to interbreed with those that become resistant, having been in a genetically modified crop. This would need to be done on a sufficiently large scale, unless there are other non-GM crops on which the pests will feed grown in the same area. The refuge area must be at least 20% of the area of the GM crop and situated adjacent to it or at least no more than half a mile away, but ideally only separated by a pathway, ditch or road. In other places, alternative ideas on managing refugia occur. Untreated areas near crops are always desirable as they provide a refuge for beneficial insects – the predators, parasitoids and pollinators. In some situations it may be difficult to have an untreated refuge area so a new strategy suggested for pink bollworm on cotton in China is to cross transgenic Bt cotton with conventional non-Bt cotton and sow second-generation seeds. This results in a random mixture of cotton plants with three quarters of the plants producing Bt protein and one quarter without it. In an 11-year study, the non-Bt plants boosted the survival of susceptible insects and delayed selection of resistance (Wan *et al*., 2017). This strategy is similar to sowing a mixture of varieties, some of which are resistant to a pathogen and decrease its spread to the more susceptible varieties within a field. Resistance to pesticides is seldom detected in beneficial species within a sprayed arable crop, although it has been reported in some orchard crops.

Insecticide Resistance Problems

Many insect pests have been shown to have developed resistance to a large number of insecticides, although not all populations of a particular species will be resistant and in many cases resistance in individual populations may be confined to a relatively small number of compounds. These include the cotton bollworm (*Helicoverpa* spp.), the diamondback moth and the Colorado beetle. This is undoubtedly due to the way farmers have quickly changed to a different insecticide product as soon as they observed inadequate control of a pest. This may not always be due to resistance being a major factor.

In India, in an area growing cotton near Guntur in the 1980s, farmers using a very cheap, manually operated sprayer frequently returned to the agrochemical shops to buy a better insecticide. Unfortunately, although the container had a different trade name, the insecticide was often identical to the product previously used or it contained a similar insecticide with the same mode of action, so control was not improved. Once the larvae, in this case *Helicoverpa armigera*, are in the second larval instar or larger, the dose required to kill them has vastly increased, simply because the weight of a larva is much greater. The quality of spraying was so poor that an adequate dose was simply not getting where needed to control the youngest larvae, so it was not surprising that the farmers could find larvae still eating the bolls.

Diamondback Moth

Diamondback moth is a pest of brassica crops, the leaves of which are very hydrophobic, thus a high volume of a diluted pesticide spray hardly leaves any deposit on the plants (Figs 7.3–7.4). Secondly, spraying down over the foliage allows pests to remain unaffected on the undersides of leaves. Lack of control led farmers to apply sprays very frequently, sometimes every two or three days. In these situations, little if anything has been done to improve the application of the insecticide in tropical areas with small farms using manually carried sprayers. Poor application has led to many different insecticides being used on brassicas within one region, so selection of resistant strains has occurred for all the insecticides used. Resistance of larvae of the diamondback moth has probably been recorded for more insecticides than any other crop pest.

Fig. 7.3. Cabbage leaf showing water droplets that have not spread on the hydrophobic surface.

Fig. 7.4. Cabbages severely damaged by diamondback moth in Malaysia.

Brown Plant Hopper

In the Philippines, and generally in south-east Asia, farmers were encouraged to spray rice crops assuming insecticides were like fertilizers and that the yield increased without any real attention to when crop loss occurred during the season or when an insecticide should be applied. At the same time, there was a trend to grow more than one irrigated crop per year, which encouraged the brown plant hopper, *Nilaparvata lugens*, to increase. The problem was that when spraying the rice plants, little of the spray was deposited on the lower part of the plants inhabited by the plant hoppers, so control was poor and natural enemies suffered. Resistance to organophosphate and carbamate insecticides was subsequently detected by the 1980s.

Colorado Beetle

Lepinotarsa decemlineata, the Colorado beetle, has spread throughout the world. It moved northwards within the USA reaching Nebraska, attacking potato crops in 1859, but reached France in 1922 and has since spread eastwards, reaching China relatively recently (1990s). It has developed resistance to over 50 different insecticides from all major insecticide classes. Following the introduction of imidacloprid, resistance was detected when control started to fail after about ten years of intensive use. The level of resistance varies among different populations and between beetle larval and adult stages, but in some situations it can be very high (up to 2000-fold). Resistance mechanisms include enhanced metabolism – involving esterases, carboxylesterases and monooxygenases – and target site insensitivity, but some of the resistant beetles show that they are less resilient when not exposed to sprays. Rotation of different insecticides and having untreated refuges have been suggested, but its control remains a major challenge. In the UK, any sighting of the arrival of Colorado beetle has to be reported, so a blitz of chemical control on the initial invaders, often arriving with shiploads of timber, has prevented their survival so far.

Peach Potato Aphid

Resistance to insecticides by the peach potato aphid *Myzus persicae*, a major global pest due to its direct damage and being a vector of viruses, has been studied extensively over 40 years. It has been shown that it has had the ability to evolve at least seven independent mechanisms of resistance that avoid or overcome the toxic effects of insecticides (Bass *et al.*, 2014). With a very short generation time, it is not surprising that resistance to an insecticide can develop rapidly. An early report of resistance to organophosphate insecticides was in 1955 (Anthon, 1955) and has continued with other groups of insecticides, as shown in Fig 7.1. In the period shown, a new group of insecticides has come in time to replace earlier insecticides, provided the aphids are not resistant to the new chemical with a different mode of action. There are some new insecticides that can control this aphid including pymetrozine, flonicamid, spirotetramat and cyantraniliprole, but there needs to be a system of greater control, perhaps on a regional basis, aided by a monitoring system to detect changes in susceptibility before it affects field control, thus maintaining susceptibility to as many different modes of action as possible.

Anopheles Mosquitoes

Since 1990 there has been an increased use of long-lasting insecticide-impregnated bed nets aimed at controlling the vector of malaria, *Anopheles* spp., inside houses, as people sleeping under a net attract the mosquitoes. In some areas, there is also indoor residual spraying. The pyrethroid insecticide impregnated into the fibres of the nets remains active for more than three years and is effective in killing mosquitoes that land on its surface and try to find a way into the person sleeping underneath. This means that mosquitoes are exposed to pyrethroid insecticide throughout the year even where mosquito populations peak during the wet season. Inevitably, mosquitoes have become resistant to pyrethroids. One possible solution has been to add piperonyl butoxide (PBO), an inhibitor of mixed-function oxidases implicated in pyrethroid resistance. This aims to kill pyrethroid-resistant mosquitoes and thus increase the efficacy of the nets. At the same time insecticides with different modes of action are being considered for use on bed nets. In 2017, a neonicotinoid and chlorfenapyr have been approved for use with bed nets. Even with this development, it seems that care is needed in the distribution of new nets being developed with other insecticides, so that within one area is one type of net that can be rotated with nets in a different area. The life of a net is about three years as tearing or washing of the net reduces the effectiveness of the insecticide.

In contrast to the situation in Africa, mosquito control in the USA is to combat the nuisance of bites, but also to minimize spread of dengue, west Nile fever and encephalitis where these diseases occur. This is done primarily with sequential space sprays from trucks or aircraft. The space sprays apply a low dosage with no residual action so selection for resistance has not been a problem. The space treatments are supported by applying larvicides. Housing, air conditioning and other factors help restrict the impact of mosquito populations. Thus if further progress is to be achieved in the tropical climates that favour mosquitoes, more consideration needs to be given to outdoor non-residual space treatments.

Control of Vector of Onchocerciasis

In 1974, the WHO began a 20-year Onchocerciasis Control Programme (OCP) of controlling blackflies (*Simulium damnosum*) in nine countries in the Sahel area of Africa, involving weekly treatment of the larval breeding sites with the OP insecticide temephos using aircraft. The OCP averted 600,000 cases of preventable blindness and made 25 million ha of land habitable and productive. The long duration of treatment was due to the period parasites could remain in humans. Inevitably, with prolonged use of one insecticide, resistance was detected, and while several other chemicals were evaluated, a change was mainly due to spraying *Bacillus thuringiensis israelensis* (Bti) across rivers so that the particulate suspension was carried downstream through the breeding site of the blackflies. Permethrin and carbosulfan were also considered possible alternatives to temephos. In 2002, the OCP was closed after 28 years as it was realized that ivermectin could be given as a tablet to kill the parasite despite the concern that the area cleared of oncho could be reinvaded by the vector. In other parts of west Africa, not included in the OCP, there has been limited control of blackflies and treatment has been confined to areas where the biting was considered too severe for development. In Cameroon, the construction of hydro-electric dams has increased the number of blackflies biting due to an increase in the volume of white water suitable for blackfly larvae. By 2014, it was also evident that the annual treatment of people with ivermectin was no longer having an adequate reduction of oncho parasites, so the dose was increased, but that was not effective; so by 2016 it was decided that control of the vector needed to be re-appraised to see if onchoceriasis could be eliminated by an integrated approach with vector control combined with mass drug distribution. From an environmental aspect, the continual treatment of areas of white water is unlikely to be sustainable, so alternative control strategies are needed.

Detailed Analysis of Resistance

Resistance to a pesticide in insects is due to different mechanisms, namely detoxification of the insecticide before it reaches the target site, changes in the sensitivity of the target site and behavioural adaption. Metabolic and target site resistance are by far the most important in terms of their frequency and the levels of resistance they can confer, while reduced penetration of an insecticide only confers low levels of resistance, although it can compound the effects of one of the two major mechanisms when they are found in the same individual.

Plants have evolved with various chemicals that act as insecticides in order to survive in a world full of herbivorous insects that could defoliate plants if there was not some protection. Food crops have been selected that tend to have minimal amounts of natural insecticide so that they are more palatable to man, but nevertheless we consume some, such as broccoli, coffee etc., that have chemicals that man can tolerate if the amount is small and can be metabolized. In the same way insects have evolved enzyme systems that enable them to survive if exposed to 'toxic' chemicals. Thus most insects have a variety of enzymes and enzyme systems to cope with many noxious chemicals in their environment (Oppenoorth, 1984). Apart from the effect of mixed function oxidases in metabolism of insecticides, resistance in some insects is due to reduced sensitivity of acetylcholinesterase. Other insects also protect themselves to some extent by reducing the penetration of insecticide into their body. This was reported by Gunning *et al.* (1991) in Australian *Helicoverpa armigera* resistant to pyrethroids and by McCaffery and Holloway (1992), who identified nerve insensitivity and delayed penetration as the primary resistance mechanisms in pyrethroid-resistant field populations of *Heliothis virescens* in the USA. Changes in behaviour of the insect can also lead to control failures; thus the mosquitoes exposed to DDT by indoor residual spraying did increase their resistance to the spray deposits, but a large proportion of the population was repelled by the deposits and escaped to bite elsewhere. Emphasis had been on the control of endophilic species, but there is now more concern that outdoor biting has increased, possibly due to exophilic species.

Resistance Monitoring

Various biological methods have been used to detect and measure resistance to insecticides and acaricides. Typically, a sample of the pest is collected and exposed to a treated surface, or individuals are treated topically using a micro-syringe, but the resistance factor can vary depending on the method used. In assessing resistance of field collected mites (*Tetranychus urticae*) to dicofol, Dennehy *et al.* (1983) showed a difference of five- to seven-fold resistance using slide-dip bioassay technique compared with 544-fold with a leaf residual bioassay, presumably due to differences in availability of the dicofol on different surfaces. The Insecticide Resistance Action Committee (IRAC) has published on their web page various IRAC Approved Test Methods to assess resistance with certain pest species. The trend now is to adopt biochemical assay methods to detect resistance in some species, thus resistance in individual aphids of the species *Myzus persicae* can be based on increased activity of carboxylesterases responsible for resistance to organophosphorus and carbamate insecticides (Needham and Sawicki, 1971), a technique that was successfully used to monitor the frequency of resistance of the aphid on outdoor crops (Sawicki *et al.*, 1978). Hemingway *et al.* (1987) have used similar methods for resistance detection in mosquito species.

As many investigators of resistance have pointed out, it is essential to detect a low level of resistance selection to avoid future difficulties in achieving control in the field, as once resistance is entrenched in the population it can rapidly reassert the effect once selection is repeated. What is needed is an insecticide resistance surveillance system (Kelly-Hope *et al.*, 2008) using molecular assays to detect resistant alleles (Ranson *et al.*, 2011), although this is costly, and a pragmatic programme to rotate modes of action to avoid pests becoming too adjusted to any one mode of action.

The exponential phase of increasing cases of resistance to fungicides and herbicides in the field occurred later than for insecticides. This may be linked to the later development of more effective products for these groups compared with insecticides, although in the case of weed control the effect of dilution by seeds germinating after the onset of resistance would have had a dilution effect.

Resistance of Pathogens to Fungicides

Until the 1970s, resistance to fungicides was insignificant; plant pathologists were even feeling confident that the problems with resistance to insecticides that were well apparent by then could never happen to them! With hindsight, it is clear that this situation only prevailed because most fungicides in use at the time were multi-site inhibitors where the multiple mutational changes that would be expressed as resistance are much less likely. The few cases where resistance has been reported, for example resistance in Pyrenophora to mercury, seem to involve reduced toxicant penetration rather than target site mutation.

Since the 1970s, virtually every fungicide group introduced has encountered resistance to some degree or other. It is no coincidence that these are largely single site inhibitors. With some fungicides, resistance, when it occurred, imparted a large shift in sensitivity and a considerable selective advantage to the newly evolved mutant. This happened most dramatically with the benzimidazoles (Smith, 1988) and the phenylamides. A similar pattern occurred with other new fungicides including the stobilurins (Heaney *et al.*, 2000). Resistance to stobilurins in *Magnaporthe grisea* (grey leaf spot) occurred very rapidly where frequent weekly sprays were applied throughout the summer on golf courses to control numerous pathogens on the highly maintained grass. Against certain pathogens on some crops, disease control was lost and alternative measures had to be deployed rapidly. In other cases, notably the DMI fungicides and the morpholines, resistance was manifest as a gradual drift towards lower sensitivity over a number of seasons. To some extent, this can be combated by raising the dose rate, although a better strategy would normally be to alternate with other chemicals and to include some other completely different disease control strategy in the cropping cycle.

The extent to which resistance can be managed depends in part on the fitness deficit exhibited by resistant mutants, which works against them when a fungicide is withdrawn. Resistance to benzimidazoles seems to be almost fitness-neutral and persists long after (decades at least) compounds such as benomyl are withdrawn. In sharp contrast, isolates of pathogens such as *B. cinerea*, resistant to dicarboximides such as iprodione, exhibit a significant fitness penalty (they can be shown to be osmotically sensitive in the laboratory) and tend to die out even within a single growing season. It is generally assumed that the wild type represents the pinnacle of fitness, being the end product of fungicide-free natural selection over evolutionary time.

Similar to other pesticide groups, the Fungicide Resistance Action Group (FRAC) (Brent and Holloman, 2007) aims to provide guidelines on fungicide resistance management and advice on the use of fungicides to reduce the risk of resistance developing and to manage it should it occur in order to limit crop losses. Advocated strategies include the greater use of varietal resistance, crop rotation, alternating between chemical groups and using mixtures of chemicals. An important development has been the use of mixtures incorporating one or more ‘modern’ systemic fungicides with at least one multi-site inhibitor from the pre-1970s arsenal of chemicals. Several of these, such as captafol (until withdrawn), mancozeb and chlorothalonil, having become, effectively, obsolete once newer chemicals reached the market, have had a renaissance of use mixed with newer chemicals as an insurance against resistance problems.

Black Sigatoka Disease

Large plantations, or many small areas of bananas in one area with one variety, suffer from Sigatoka disease (*Mycosphaerella musicola*) and, more recently, black Sigatoka disease *Mycosphaerella fijiensis* var. *difformis*. Attempts to protect the new young leaves are difficult with ground equipment as lower leaves are a barrier to spray reaching the vertical unfurled leaf, which is most susceptible to the disease. Sprays in most major banana production areas are applied using aircraft. Frequent fungicide sprays have led to resistance being detected, so special recommendations have been given by the FRAC for bananas in which a maximum number of applications is set for each mixture and there are restrictions on timing of applications. As an example, the total number of sprays with fungicides that are succinate dehydrogenase inhibitors (SDHIs), such as fluopyram, fluxapyroxad and isopyrazam, used in a mixture, should not exceed three per season and should be not more than a third of the total number of sprays applied, the initial treatment being at the start of the onset of the annual period of disease and subsequent treatments separated by at least three months of an SDHI-free period.

Rice Blast

A major disease on rice is blast caused by the fungus *Magnaporthe oryzae*, although it used to be kept in check by growing varieties resistant to the fungus. It can affect all above-ground parts of a rice plant: leaf, collar, node, neck, parts of the panicle and the leaf sheath, killing seedlings or plants up to the tillering stage. At later growth stages, a severe leaf blast infection reduces leaf area for grain fill, reducing grain yield. There has therefore been an increase in fungicide applications, but the disease is less serious where soils have a high silica content, so the greatest reduction on blast incidence was observed where 4 g silica/l of spray were applied, regardless of the pH of the soil.

Septoria Blotch on Wheat

Fungicides are applied to control the pathogen *Zymoseptoria tritici* as it is a global threat to sustainable wheat production. DMI fungicides have been the main type of fungicide used, but resistance problems are challenging their use. Heick *et al.* (2017) have recently examined fungicide spray strategies using one, two or three applications, alternations or mixtures of different DMIs and DMIs mixed with other modes of action including a **s**uccinate **deh**ydrogenase **i**nhibitor (SDHI) and a multi-site inhibitor. Their best yield results and control were attained by a diversified DMI strategy, which also included the SDHI boscalid and the multi-site inhibitor folpet. They indicated that selection of resistance was less with a more diversified strategy, and encouraged the adoption of mixing and alternating fungicides into spray strategies to minimize the risk of resistance build-up and to prolong the effective life of fungicides.

Herbicide Resistance Problems

Herbicides have been most widely used since the 1950s with the detection of a triazine-resistant weed (*Senecio vulgaris*) in 1968 (Ryan, 1970). There are now well over 200 weed species that have become resistant to them, many with multiple resistance to herbicides with different modes of action (see Table 4.1 to see different modes of action of herbicides). As with insecticides and fungicides, it is the continual exposure of weeds to herbicides with the same mode of action that inevitably leads to selection of resistant weeds. Initially, the presence of weeds resistant to the triazines did not cause much concern as resistant biotypes were easily controlled by specific alternative herbicides (Shaner, 2014).

The problem has become most acute where glyphosate has been applied in a range of crops, which are genetically engineered to be tolerant of glyphosate. In the USA, genetically modified (GM) crops tolerant to herbicides have been widely grown since 1996, with farmers benefiting by additional farm income, which is estimated to be about $21 billion, and at the same time they used 225 million kg less herbicide-active ingredient between 1996 and 2012 (Brookes, 2014). The downside is that weeds resistant to glyphosate now occupy up to 4 million ha and are being referred to as 'superweeds'. The development of a herbicide-resistant crop enabling the use of a particular herbicide was initially welcomed by farmers, as they could delay a treatment without an adverse effect on the crop. However, treating vast areas with one chemical, glyphosate, made the evolution of superweeds inevitable. Industry was already beginning to respond by developing crops with tolerance to other herbicides, notably 2,4-D and dicamba. Trials showed that dicamba is an effective alternative mode of action to glyphosate in fields where the glyphosate-resistant weed Palmer amaranth occurs (Inman *et al.*, 2016). However, these herbicides are volatile, with vapour drift affecting susceptible crops downwind (Egan *et al.*, 2014); so new formulations have been developed to minimize this. At the same time, farmers have added other herbicides with different modes of action with glyphosate, even where instances of weed resistance to glyphosate have not been found, to maintain effective weed control using no-till and conservation tillage.

Black Grass

In northern Europe, black grass (*Alopecurus myosuroides*) is one of the most important weeds. It is an annual weed with round, slender stems that can grow up to 90 cm high. The seeds were usually buried to a sufficient depth to prevent more than a few instances of black grass in a subsequent season, but sowing wheat in the autumn with no real break between crops has led to germination of black grass seeds and increasing difficulty in controlling the weed with herbicides. Distribution of herbicides with the closely spaced wheat plant no doubt deposited too little on the weeds, and ultimately, multiple resistance was detected to all the herbicides used. A further problem is that some herbicides cannot be used where there is a risk of contamination of ground water.

Rather than ploughing prior to sowing winter cereals, which can significantly reduce black grass, some farmers have adopted minimum tillage to cause the weed seeds on the soil surface to germinate quickly and then get dried out, before sowing later. Adopting a rotation with spring-sown wheat or an alternative crop enabled a break that can reduce black grass better compared with autumn sowing. With wider row spacing of other crops, alternative herbicides can be directed more effectively on weeds in the inter-row (Fig 7.5). The effects due to different cultural practices will vary depending on climatic conditions, so farmers

Fig. 7.5. No-drift herbicide application in inter-row application. (Photo courtesy of Micron.)

need help if they are to manage their weeds better and avoid resistance to the herbicides they can use.

Pigweed

One of the superweeds is *Amaranthus palmeri*, which has been foraged for a long time by the native American population. It can grow rapidly and with a strong taproot can penetrate hard soil to reach water and nutrients not available to shallow-rooted crops. Pigweed has a long history of becoming resistant to herbicides, with trifluralin in the 1980s and then imazaquin in the 1990s, and now glyphosate. A survey in North Carolina in 2005 showed that nearly 20% of 290 populations sampled were resistant to glyphosate with resistance to ALS-inhibiting herbicides nearly universal, but resistance to fomesafen or glufosinate was not observed (Poirier *et al.*, 2014). With dicamba-tolerant crops, resistance to dicamba will no doubt occur. Similar problems have been recorded in other countries with a shift to more annual broad-leaved weeds being resistant to herbicides (Johnson *et al.*, 2009).

Early on, with resistance to DDT detected, Winteringham at the Pest Infestation Laboratory in Slough recognized in the 1960s that problems of resistance, which had already become more intense and widespread, would continue and that onset of resistance may be delayed by avoiding all forms of unnecessary selection pressure, such as using persistent chemicals. He made the important point of chronological and geographical restriction of chemical applications to keep pest populations at economic injury levels (Fletcher, 1974). Sadly, little recognition of this has been taken, so without management of different modes of action at a national level resistance has continued to be a major problem.

The Way Forward

While the past has seen hope for a new molecule with a new mode of action, the need is for greater cooperation over large areas to restrict use of one pesticide in a single season and endeavour to rotate use of different modes of action to limit the selection pressure for resistance within each cropping season. In vector control, dividing a country into distinct areas and rotating insecticides with different modes of action, whether applied as sprays or impregnated in bed nets, are needed to sustain effective vector control.

Forrester (1990) stated that preventative insecticide resistance management (IRM) is preferable to adopting a curative approach, which has a lower chance of long-term success. As pointed out by Onsted (2008), the best IRM programme will take advantage of implementation of the best integrated pest management practices.

References

Anthon, E.W. (1955) Evidence for green peach aphid resistance to organophosphorous insecticides. *Journal of Economic Entomology* 48, 56–57.

Babers, F.H. (1949) *Development of Insect Resistance to Insecticides.* USDA Agriculture Research Administration Bureau of Entomology and Plant Quarantine, Washington, D.C.

Bass, C., Puinean, A.M., Zimmer, C.T., Denholm, I., Field, L.M. *et al.* (2014) The evolution of insecticide resistance in the peach potato aphid, *Myzus persicae. Insect Biochemistry and Molecular Biology* 51, 41–51.

Brent, K.J. and Holloman, D.W. (2007) Fungicide resistance in crop pathogens: How can it be managed? *FRAC Monograph No. 1.* Fungicide Resistance Action Committee, Basel, Switzerland.

Brookes, G. (2014) Weed control changes and genetically modified herbicide tolerant crops in the USA, 1996–2012. *GM Crops Food* 5, 321–332.

Debach, P. and Bartlett, P. (1951) Effects of insecticides on biological control of insect pests of citrus. *Journal of Economic Entomology* 44, 372–383.

Dennehy, T.J., Granett, J. and Leigh, T.F. (1983) Relevance of slide-dip and residual bioassay comparisons to detection of resistance in spider mites. *Journal of Economic Entomology* 76, 1225–1230.

Duncombe, W.C. (1973) The acaricide spray rotation for cotton. *Rhodesian Agricultural Journal* 70, 115–118.

Egan, J.F., Barlow, K.M. and Mortensen, D.A. (2014) A meta-analysis on the effects of 2,4-D and Dicamba drift on soybean and cotton. *Weed Science* 62, 193–206.

Fletcher, W.W. (1974) *The Pest War.* Blackwell, Oxford, UK.

Forgash, A.J. (1984) History, evolution and consequences of insecticide resistance. *Pesticide Biochemistry and Physiology* 22, 178–186.

Forrester, N.W. (1990) Designing, implementing and servicing an insecticide resistance management strategy. *Pesticide Science* 28, 167–179.

Gunning, R.V., Easton, C.S., Balfe, M.E. and Ferris, I.G. (1991) Pyrethroid resistance mechanisms in Australian *Helicoverpa armigera. Pesticide Science* 33, 472–490.

Heaney, S.P., Hall, A.A., Davis, S.A. and Olaya, G. (2000) Resistance to fungicides in the QoI-STAR cross-resistance group: current perspectives. *Proceedings of British Crop Protection Conference – Pests & Diseases*, 755–762.

Heick, T.M., Justesen, A.F. and Jorgensen, L.N. (2017) Anti-resistance strategies for fungicides against wheat pathogen *Zymoseptoria tritici* with focus on DMI fungicides. *Crop Protection* 99, 108–117.

Hemingway, J., Jayawardena, K.G.I., Weerasinghe, I. and Hearath, P.R.J. (1987) The use of biochemical tests to identify multiple insecticide resistance mechanisms in field-selected populations of *Anopheles subpictus* Grassi (Diptera: Culicidae). *Bulletin of Entomological Research* 77, 57–66.

Inman, M.D., Jordan, D.L., York, A.C., Jennings, K.M., Monks, D.W. *et al.* (2016) Long-term management of Palmer Amaranth (*Amaranthus palmeri*) in dicamba-tolerant cotton. *Weed Science* 64, 161–169.

Johnson, W.G., Davis, V.M., Kruger, G.R. and Weller, S.C. (2009) Influence of glyphosate-resistant cropping systems on weed species shifts and glyphosate-resistant weed populations. *European Journal of Agronomy* 31, 162–172.

Kelly-Hope, L., Ranson, H. and Hemmingway, J. (2008) Lessons from the past: managing insecticide resistance in malaria control and eradication programmes. *Lancet – Infection* 8, 387–389.

McCaffery, A.R. and Holloway, J.W. (1992) Identification of mechanisms of resistance in larvae of the tobacco budworm *Heliothis virescens* from cotton field populations.

Proceedings of the Brighton Crop Protection Conference – Pests and Diseases – 1992, 1, 227–232. BCPC, Farnham, UK.
Melander, A.L. (1914) Can insects become resistant to sprays? *Journal of Economic Entomology* 7(2), 167–173.
Needham, P.H. and Sawicki, R.M. (1971) Diagnosis of resistance to organophosphorus insecticides in *Myzus persicae* (Sulz.). *Nature* 230, 125–126.
Onsted, D.W. (2008) The future of insect resistance management. In: Onsted, D.W. (ed.) *Insect Resistance Management – Biology, Economics and Prediction*. Academic Press, Amsterdam.
Oppenoorth, F.J. (1984) Biochemistry of insecticide resistance. *Pesticide Biochemistry and Physiology* 22, 187–193.
Poirier, A.H., York, A.C., Jordan, D.L., Chandi, A., Everman, W.J. and Whitaker, J.R. (2014) Distribution of glyphosate- and thifensulfuron-resistant Palmer amaranth (*Amaranthus palmeri*) in North Carolina. *International Journal of Agronomy* 2014: 10.1155/2014/747810.
Ranson, H., N'Guessan, R., Lines, J., Moiroux, N., Nkuni, Z. and Corbel, V. (2011) Pyrethroid resistance in African anopheline mosquitoes: What are the implications for malaria control? *Trends in Parasitology* 27, 91–98.
Ripper, W.E. (1956) Effect of pesticides on balance of arthropod populations. *Annual Review of Entomology* 1, 403–408.
Ryan, G.F. (1970) Resistance of common groundsel to simazine and atrazine. *Weed Science* 18, 614–616.
Sawicki, R.M., Devonshire, A.L., Rice, A.D., Moores, G.D., Petzing, S.M. and Cameron, A. (1978) The detection and distribution of organophosphorus and carbamate insecticide-resistant *Myzus persicae* (Sulz.) in Britain in 1976. *Pesticide Science* 9, 189–201.
Shaner, D.L. (2014) Lessons learned from the history of herbicide resistance. *Weed Science* 62, 427–431.
Smith, C.M. (1988) History of benzimidazole use and resistance. In: Delp, C.J. (ed.) *Fungicide Resistance in North America*. American Phytopathological Society, St Paul, Minnesota, pp. 23–24.
Wan, P., Xu, D., Cong, S., Jiang, Y. and Huang, Y. (2017) Hybridizing transgenic Bt cotton with non-Bt cotton counters resistance in pink bollworm. *Proceedings of the National Academy of Science*, 114(21), 5413–5418.

8 Integrated Pest Management

The early use of the term integrated pest management (IPM) was attributed to Stern *et al.* (1959) when they introduced the concept in a presentation on integration of chemical and biological control of the spotted alfalfa aphid (*Therioaphis maculata*). Their aim was the selective use of a lower dosage of a broad-spectrum insecticide to give some protection to natural enemies, coupled with linking pesticide application to economic thresholds. However, Kogan (1998) attributes the idea to Hoskins *et al.* (1939) who already realized that there should be more discrimination in using insecticides. A little earlier, Ripper *et al.* (1951), noting the selection of pest populations resistant to the use of insecticides and resurgence of pests in the absence of natural enemies, proposed the supplementation of biological control with chemical control by using selective insecticides. This could be achieved by finding chemicals that allowed beneficial insects to survive or by a judicious choice of the right concentration and application technique. Sufficient pests survived to maintain an effective density of natural enemies.

Setting out to control cotton bollworms in Africa, Tunstall *et al.* (1959) concluded their assessment of different methods of control by stating that it was important to establish a more economic spraying routine in conjunction with other methods of control. This led to using a closed season of at least two months prior to sowing the cotton variety Albar, with resistance to jassids (*Jacobiasca fascialis*) due to pubescence (Figs 8.1, 8.2) and the disease bacterial blight *Xanthomonas gossypii*. This was followed by monitoring the crop by regular bollworm egg counts to determine when to apply insecticides, if needed, between weeks 8 and 16 after germination of the seeds (Fig. 8.3). The threshold for deciding when to spray against *Helicoverpa armigera* was an average of half an egg per plant based on sampling 12 plants across the diagonals of the field (Fig. 8.4). The hairiness of the Albar variety for jassid resistance aided the deposition of sprays,

Fig. 8.1. A glabrous variety showing jassid damage with resistant plant in adjacent row. Unsprayed trial in Malawi 1969.

Fig. 8.2. Hairy leaf of cotton variety that deters infestation by jassids.

as the hairs on leaves and stems collected insecticide particles and also increased the time first instar larvae took to find a bud or boll. Control of the young larvae was important before they became larger (Fig. 8.5). After harvesting, the crop residues were collected and burnt on small farms, or shredded (Fig. 8.6). Ploughing eliminated any food source for cotton pests before the next rainfall season and ratooning was banned.

Between the availability of DDT and the first mention of integrated pest management, there was, according to Metcalf (1980), the Age of Optimism when it was thought that pesticides would solve all pest problems, but Rachel Carson's book *Silent Spring*, highlighting the dangers of wide-scale and uncontrolled use of pesticides, initiated an Age of Doubt

Fig. 8.3. Pegboard used to scout and determine population of red (*Diparopsis castanea*) and American bollworm (*Helicoverpa armigera*) eggs.

Fig. 8.4. *Helicoverpa armigera* egg.

(1962–1976). Pimentel *et al.* (1978) estimated that despite the use of pesticides, the cost of which had risen due to the high cost of development and getting products registered, crop losses from insect attack had increased from 7% in the 1940s to 13%, but this is against a background of changing agricultural practices and improved varieties, and better weed control and soil fertility. Nevertheless, as DDT was relatively inexpensive, it was used at high rates and mixed with toxaphene on cotton as mentioned in Chapter 1. With aerial spraying, deposition within the crop canopy was not always as good as it should have been, especially if a proportion of the spray was lost in thermal airflows and downwind drift. Nevertheless, the

Fig. 8.5. *Helicoverpa* larva on boll.

Fig. 8.6. Shredding cotton stalks after harvest.

benefit of applying insecticides was estimated at $2 or more for each $1 spent on control.

This period was when the sole reliance on chemical control, especially with broad-spectrum insecticides, was gradually recognized as a 'no-win' strategy. The combined interaction of insect pest resistance, pest resurgences and outbreaks of secondary pests provided a positive feedback that defeats the strategy of chemical control, so the need for a rational and sensible use of pesticides was recognized (Luck *et al.*, 1977). Ripper (1956) had pointed out that to avoid an unfavourable impact of pesticides on the balance of arthropod populations, consideration needed to be given to using non-selective chemicals by timing an application with a lower dosage to minimize the effect on non-target arthropods or to develop a selective pesticide that is specific to the pest without injuring the natural enemies. Unfortunately, few pesticides have sufficient selectivity, although pirimicarb, which was marketed for aphid control and application of baculoviruses, has been successful in some situations.

The latest recognition of IPM is within the European Regulations, which have included a policy of IPM within the Sustainable Use of Pesticides Directive 91/414/EEC, and EC Regulation 1107/2009, which came into force in June 2011. This requires pesticide use to be considered from three aspects: economic, social and environmental. The environmental aspect is particularly important as it takes account of Rachel Carson's prediction of a silent spring without birds singing, and a growing public disquiet with current practices. In the five decades since her book was published there has been significant progress achieved by various conventions, notably the Stockholm Convention banning the application of persistent organic pollutants, DDT being a key factor. Studies at the time were showing that these chemicals, moving through the food chain, had a major impact on birds' egg shells, making them thinner and liable to break easily. While pesticides have continued to be blamed for the reduction in the population of many bird species, changing farming practices to maximize profitable food production have led to loss of habitat, with larger fields losing hedgerows and unfavourable methods of harvesting. Nevertheless, the forecast of a 'silent spring' has been a major influence encouraging much improved pesticide management.

Recognition of the dangers of using highly hazardous (toxic) pesticides (HHPs) and the need for training users, who had to wear protective clothing, has gradually led registration authorities to withdraw the use of many HHPs and to support more research on finding less toxic alternative pesticides. Concerns about spray drift have also led to changes in application techniques and the use of unsprayed buffer zones. No doubt further changes will occur, especially to avoid polluting water supplies, as more emphasis is given to environmental protection.

IPM Defined

Over the last 50-plus years, there have been many different views expressed as to what IPM actually means. Perhaps the biggest contrast was that IPM aimed to avoid using any synthetic pesticides and relied entirely on biological control by using resistant crop varieties, predators, parasitoids and, if necessary, 'botanical' (i.e. natural) insecticides and cultural practices. This was championed by those growing 'organic' crops and preferring organic foods, even though copper sprays were used, encouraged by the Soil Association. Following the views of Stern *et al.*, the emphasis was on crop monitoring so that the timing of pesticide applications was related to an economic threshold. This was taken up by those who were specialists in developing models to determine whether a programme of treatments could provide an economic return for the farmer. They referred to the problem of controlling Colorado beetle (*Leptinotarsa decimlineata*) on potatoes, which had been of minor importance until about 1850, when it spread north as Americans started to grow potatoes on a much greater scale and started to apply an arsenate to control it (Fig. 8.7). Later they added copper/lime-based Bordeaux mixture to control late blight.

In introducing the concept of IPM, Stern *et al.* (1959) argued that chemical control should be employed to reduce a pest population that rises to a dangerous level when natural controls are inadequate. When such a pesticide is used, the cost must cover not only the amount lost due to pest damage but also the possible deleterious effects on the ecosystem. Essentially, chemical control should only be used when natural controls are inadequate and should act as a complement to biological control. As a chemical 'spray' cannot search out the pest like a predator, application is crucial so that the pesticide is deposited where the pest is located or translocated there within the treated plants. Plant growth and loss of spray deposits due to weathering means that in many situations the chemical treatment cannot restrain an increase in pest abundance without

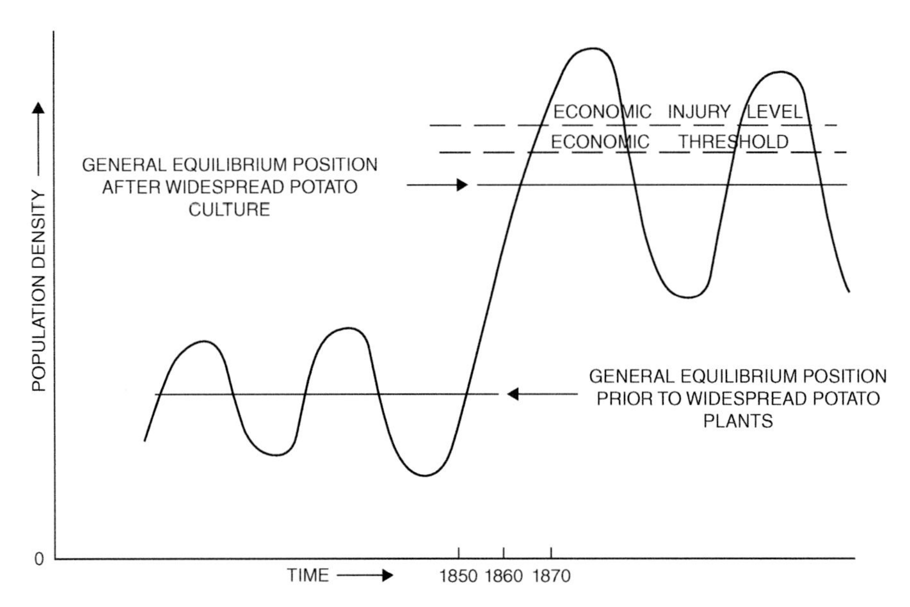

Fig. 8.7. Schematic graph of the change in general equilibrium position of the Colorado beetle following widespread growing of potatoes in the USA. (From Stern *et al.*, 1959)

repeated applications. Thus, sprays may have to be repeated at varying intervals of time, so routine monitoring of a crop is needed to detect when the pest population has returned to a level that justifies another spray. Stern and colleagues also advocated that consideration had to be given to the insects, diseases, plant nutrition, plant physiology and plant resistance, as well as the economics of the crops. Today, weed management and having refuge areas of flowering plants is being recognized as important to protect natural enemies and pollinators.

The Components of IPM

The term integrated pest management as defined by Stern *et al.* (1959) related specifically to an improved timing of insecticide applications, so it was, essentially, integration of chemical and biological control of insects (Van Den Bosch and Stern, 1962) rather than a broader view of pests to include diseases and weeds. The emphasis in the cotton example, mentioned in Chapter 1, was on using insecticides with cultural and biological controls. Some refer to this as crop pest management or simply pest management, the latter referring to a more area-wide or regional adoption, rather than by individual farmers. Too often in modern society the systems used depend on the financial world, namely which crops enable the farmers to achieve the best return on their investment. This often makes it difficult to retain the traditional rotation of crops that were largely developed for facilitating weed management and by using a legume in the rotation, to help maintain soil fertility. Clearly, crop rotation remains an important part of pest management, even if herbicides can effectively keep weed populations in check, as overuse of a particular herbicide will inevitably select weeds resistant to the herbicide used in a particular crop. Crop rotation also helps to avoid a build-up of soil pests and maintain soil fertility, a factor that has, to a large extent, been forgotten, with modern agriculture requiring applications of fertilizer. Nevertheless, perhaps due to problems of weeds resistant to certain herbicides, some farmers are now adopting rotation of crops. Thus, one farmer in the UK has a seven-year rotation involving a soil management programme that allows a different weed management programme (Table 8.1).

In colder parts of the world, the winter provided a good break from pests, but with climate change and less severe weather, the dominance of

Table 8.1. A rotation scheme to reduce the continuity of sowing winter wheat too frequently.

Year	1	2	3	4	5	6	7
Planted crop	Winter barley	Winter oilseed rape	Winter wheat	Sugar beet	Winter wheat or spring barley	Spring beans	Winter wheat

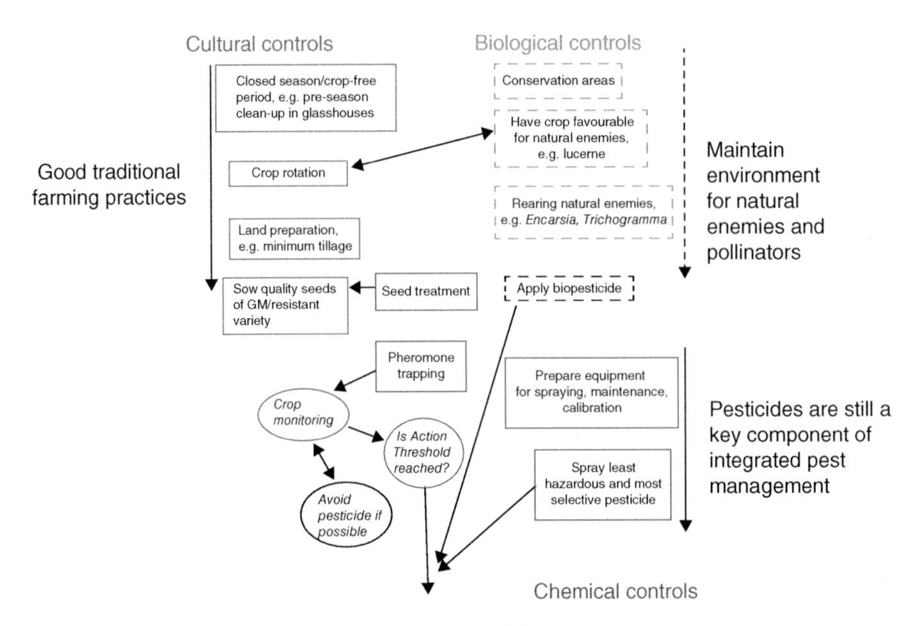

Fig. 8.8. Diagram to show the components of IPM.

autumn-sown wheat without proper rotation of crops was undoubtedly a factor in black grass (*Alopecurus myosuroides*) becoming a 'superweed', requiring a change in approach to its control. In the tropical countries, having a closed season is particularly difficult where irrigation is available to grow some crops, such as tomatoes, all year round. Build-up of diseases such as blight in the wet season inevitably transfers to dry-season crops unless there is a closed season and sufficient spatial separation and rotation of crops.

While ploughing provided some weed management of annual crops, there is an adverse effect on some soil fauna, such as earthworms. With increasing use of pesticides, soil compaction has become a more important issue with the weight of large volumes of water being used in spray applications. Thus while occasional ploughing may still be essential, minimum tillage is now an important component of IPM, although much depends on the crops being grown and whether they can develop good root systems.

Plant Resistance to Pests

Perhaps the key 'cultural' control is the choice of crop variety, which no doubt will have to adapt from traditional plant breeding techniques to utilize the scientific development of genetically engineered varieties to express resistance to pathogens and insect pests. A glabrous variety, Cokers wild, from the USA, was sown when the crop was introduced into Southern Rhodesia in 1924, but success of cotton as a crop in southern

Africa depended initially on the plant breeders in South Africa who introduced pubescent varieties from India to select 9L34, a variety that survived jassid infestations (Parnell *et al.*, 1949). Then, the crop was essentially grown 'organically' with the area sown dependent on whether there was a price incentive or whether yields had been slightly better the previous year. This continued until 1958 when a further refinement was the adding of resistance to bacterial blight (*Xanthomonas malvacearum*) by growing Albar cotton to retain a high potential yield that was worth protecting from the bollworms. Sowing this variety meant that insecticide sprays were unnecessary for the first six to eight weeks after seed germination, prior to production of flower buds (Figs 8.9–8.11). It was the introduction of insecticides that marked a major increase in yields.

Some cotton varieties, especially with *Gossypium barbadense* cotton, had higher levels of gossypol that gave some protection from bollworm damage. The sowing of a resistant variety was assisted by the removal of the fuzz on cotton seeds using sulfuric acid. This enabled seeds to be sown mechanically. The seed treatment also protected young seedlings from soil-borne pathogens. Today, the new GM varieties can provide

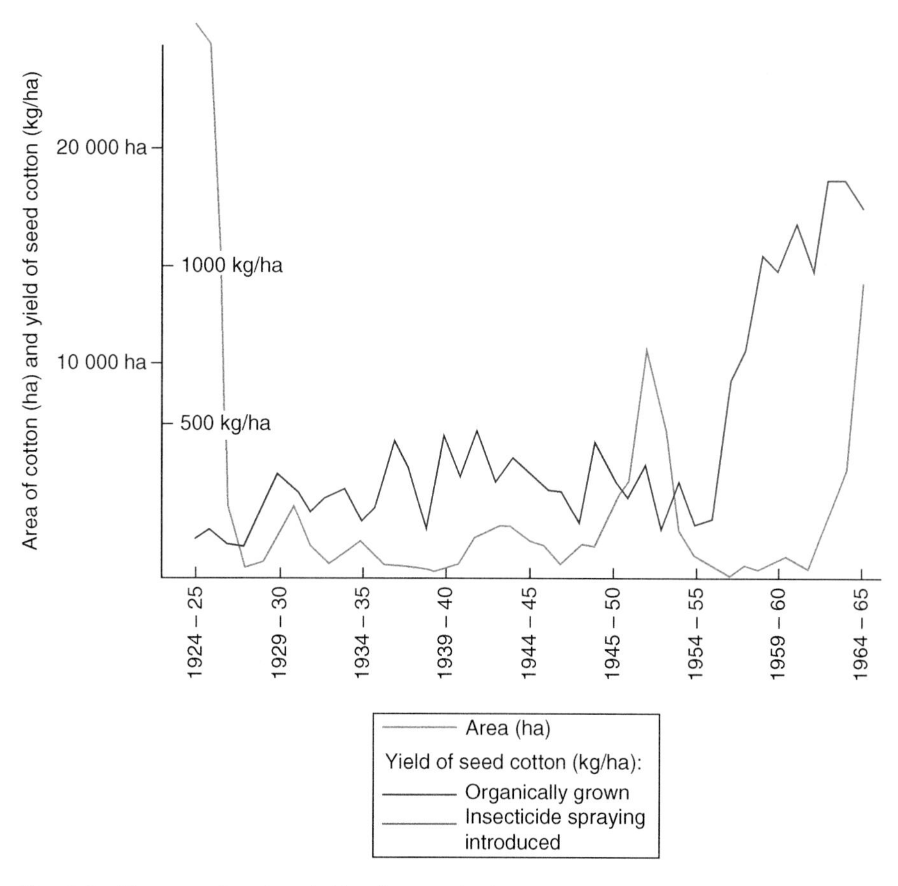

Fig. 8.9. Diagram showing yields of cotton in Southern Rhodesia from 1924 to 1965.

Fig. 8.10. Edge of an area of cotton in Malawi *c.*1971 showing unsprayed cotton in the foreground and open bolls of sprayed cotton ready for harvesting.

Fig. 8.11. Cotton bolls and buds on a branch.

tolerance to herbicides and include an integral toxin from *Bacillus thuringiensis* (Bt) to control the first instar larval stage of bollworms as soon as they start to feed on a bud or boll. However, Bt cotton is not effective against insect-sucking pests, so the host plant resistance of pubescent varieties is still important to control jassids, together with seed treatment with a systemic insecticide to control aphids and whiteflies.

Similar scenarios can be shown in other crops where plant breeders have selected varieties adapted to local conditions to yield well due to some resistance to important pathogens. In the UK in 1815, Thomas Knight noticed that some cereal plants were affected by fungi, while others were immune, so he suggested that it would be good to select the unaffected plants and develop new strains with resistance to the diseases. Plant breeders subsequently developed a number of different varieties of wheat and other cereals with different levels of resistance to the main diseases so that farmers could choose the variety most suited to their area. Wheat varieties recommended by the National Institute of Agricultural Botany (NIAB) around 1998 (Table 8.2) show that it is very difficult to get a variety with resistance to all potential disease problems, and that use of a fungicide was still needed to improve yields. However, there are new technologies in the 21st century for improving the resistance to pests and diseases, by utilizing genetic techniques to engineer plants expressing RNAi to silence multiple genes and enhance a plant's resistance to fungal and viral diseases.

Van Emden (1972) differentiated nine mechanisms by which varietal control can operate against insects (Fig. 8.12), with more than one mechanism possibly operating in a resistant variety. With most crops attacked by several pests, varietal control of at least one or two pests is essential to reduce the reliance on too many pesticides and enhance biological control.

Crop Rotation

Crop rotation is one of the oldest and most widespread systems to reduce pest populations. In Roman times, farmers followed a cropping system called 'food, feed and fallow' – namely wheat followed by barley or oats to feed cattle, and fallow to rest the soil. In many parts of the world, farmers opened up an area to cultivate but later abandoned it as weeds were impossible to control – a system called 'slash and burn'. In Europe, when potatoes were introduced from the Americas, it was noticed that soil pests increased if the farmer did not rotate crops. In the UK, in the 17th century, they adopted a rotation of turnips; a nitrogen-fixing crop – red clover; potatoes; and a cereal – wheat or barley. Likewise, in the southern states of the USA farmers were advised to rotate their cotton with soil-enriching legumes such as groundnuts and peas. In Africa, in the 1960s, cotton was followed by maize, which benefited following a deep-rooted crop, and this was followed by a legume.

Interference Methods

Pheromones

At much the same time that pesticides were being used by more farmers, scientists had discovered that the males of certain insects emitted a chemical known as a pheromone to attract their mate. Studies revealed that some species can detect and follow a pheromone trail for miles. Large water traps baited with virgin female *Diparopsis* moths attracted as many as 1000 male moths in a single trap (Fig. 8.13). Sending pupae to London enabled the components of the pheromone to be subsequently determined

Table 8.2. Choice of wheat varieties in the UK depending on their resistance to diseases.

	Reaper	Hussar	Rialto	Consort	Riband	Hereward	Spark	Buster	Soissons
*With fungicide	102	102	102	102	100	93	93	100	93
*No fungicide	88	86	86	82	75	80	80	84	82
#Mildew	7	7	7	6	7	6	7	5	8
Yellow rust	4	5	6	8	6	6	7	9	7
Brown rust	9	9	4	6	4	7	8	3	2
Septoria nodorum	7	6	5	6	5	6	7	7	4
Septoria tritici	6	5	5	4	3	6	7	5	6
Fusarium ear blight	5	6	5	7	6	6	7	6	5
Eyespot	5	5	6	5	5	5	5	6	4

*Yield as percentage of a treated control (9.73 t/ha)
#Scale 1–9; highest value shows character to a high degree.

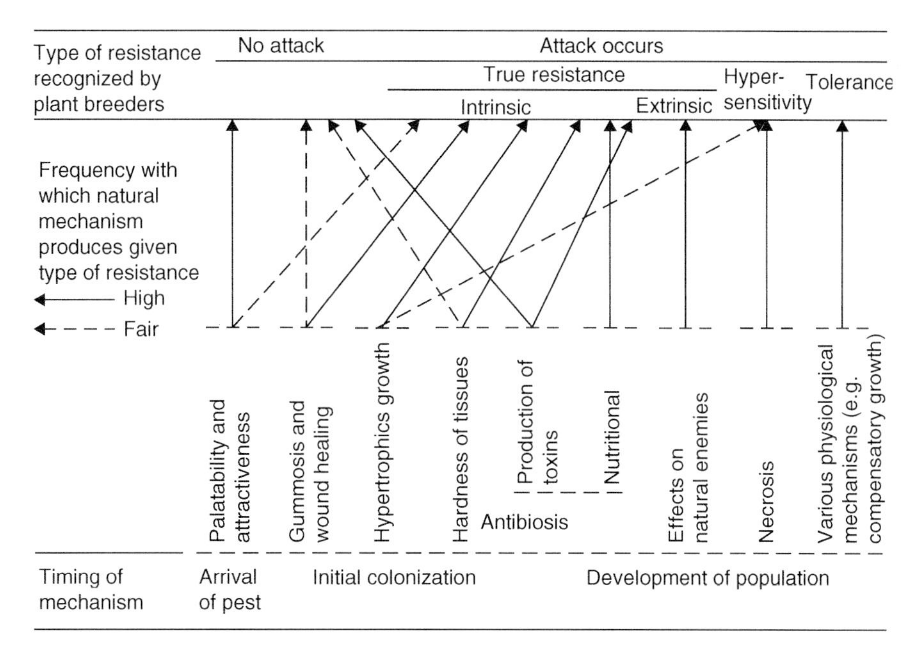

Fig. 8.12. Nine ways in which plants can show resistance to insects. (Redrawn from from Van Emden, 1972.)

Fig. 8.13. Water trap with virgin *Diparopsis* moths in cage above the water. A single trap can catch hundreds of moths in a single night.

and synthesized (Nesbitt *et al.*, 1975), but the following cage and field studies (Marks, 1976; Marks *et al.*, 1978) did not result in it being used either in traps or as a ULV spray to confuse the natural population and reduce oviposition. The unanswered question is whether it would have been sufficiently effective in the absence of an insecticide spray programme, but it has not been tried in the context of organic cotton.

Elsewhere, the pheromone of the pink bollworm *Pectinophora gossypiella* – gossyplure was successfully used as a micro-encapsulated formulation at the rate of 10 g/ha and gave results comparable to those of chemical insecticides. In 1994, in Egypt, the pheromone was applied on 150,000 ha of cotton, 45% of the national crop. The principal method of using it was with a twist-tie (Fig. 8.14) that was attached at intervals by hand through the crop, but a micro-encapsulated formulation was also tried and applied aerially. Successful control of pink bollworm has been reported from the USA and China. Pheromone traps have been used to monitor various other pests, including the boll weevil *Anthonomus grandis* and corn earworm *Helicoverpa zea* (Fig. 8.15). Use of pheromones has also been in stored-product insect control and with some pests in glasshouses.

The highly selective attraction of the pheromone has been used in traps to monitor populations and, combined with an insecticide, used as an 'attract and kill' technique. Olive fly (*Bactrocera oleae*) control was traditionally carried out using an insecticide, usually dimethoate, applied as a patch sprayed on trees through an olive grove. McPhail traps (Fig. 8.16) were used to check fly populations. Spraying an insecticide alone against larvae proved disastrous, resulting in fewer natural enemies. Experiments were carried out with coloured sticky traps, but these also attracted beneficial insects. Since the late 1990s, the 'ecotrap' has been used for the mass trapping of the olive fruit fly. The trap is baited with the pheromone of the olive fruit fly, ammonium bicarbonate and deltamethrin. Using the pheromone as a selective control technique is particularly important where conservation of natural enemies is another component of an IPM programme. The use of pheromones in IPM was reviewed by Witzgall *et al.* (2010).

Fig. 8.14. Twist and tie pheromone dispenser on cotton in Egypt.

Fig. 8.15. *Helicoverpa* trap in the USA.

Fig. 8.16. McPhail trap in Greece to collect olive flies (*Dacus*).

'Push–pull'

A technique was developed to integrate intercropping among subsistence farmers in Africa by using a stimulus to repel an insect pest from a crop ('push') and a nearby trap crop to attract it ('pull') (Hassanali *et al.*, 2008). This system involves intercropping maize with a stemborer moth-repellent forage legume, silverleaf *Desmodium uncinatum* (push), and planting an attractive trap crop, Napier grass (pull), around the intercrop. Additionally, chemicals produced from *Desmodium* roots inhibit African witchweed (*Striga*). The system worked well where rainfall was adequate through most of the year and allowed different crops to be grown, including legumes, which improved soil fertility when rotated with a cereal crop such as maize. In areas with a long dry period, it seems to have been less successful. Getting *Desmodium* seeds to germinate and get established before the rains was identified as a problem (Edwards, 2015); however, Eigenbrode *et al.* (2016) point out that there have been a few successful applications in pest management and suggest that push–pull systems could be modified by allowing effective monitoring of pests to increase the effectiveness of the technique in IPM. Nevertheless, during the invasion of fall armyworm (*Spodoptera frugiperda*) in 2017, the maize in push–pull plots gave higher yields than maize grown without the legume intercrop (Midega *et al.*, 2018). Recognizing the value of a push–pull strategy in conservation of natural enemies, the technique is also being tried in India (Bhattacharyya, 2017).

Insect Growth Regulators

Insect growth regulators, such as diflbenzuron, provide an alternative mode of action to the insecticides that affect the nervous system. They are relatively safe to adult predators and parasites as well as all stages of predaceous mites. Unfortunately, the larval stages of some beneficial insects, such as coccinellid larvae, are affected. The main problem for users is that the speed of action is slow, particularly when sprayed against locust hoppers, but from an environmental aspect their use is preferred, except where aquatic arthropods can be adversely affected. Methoprene, mimicking a juvenile hormone, has been used successfully against mosquito larvae.

Anti-feedants and Repellents

While anti-feedants and repellents can have a localized effect, the insects repelled can move to a different area, which limits their overall impact.

Genetic Manipulation

Edward Knipling, with his colleague Raymond Bushland, received the 1992 World Food Prize for their collaborative achievements in developing the sterile insect technique (SIT) for eradicating or suppressing the threat posed by pests to the livestock and crops that contribute to the world's food supply (Knipling, 1979). Apart from SIT, Knipling also contributed to various other techniques aimed at total insect population management. The most successful implementation of a genetic manipulation has been the release of sterile screw worm flies – *Cochliomyia hominovorax* – an obligatory parasite of livestock in southern USA, central and South America. Large populations were produced on a factory scale and the pupae irradiated, so the insects were not genetically modified. The continuous culture of *Cochlomyia* can result in a decline in fitness and sexual competitiveness, so the culture periodically needed an injection of wild stock. After initial trials on the isolated island of Curacao, releases from Florida, started in 1958, were progressively moved westwards and south to Panama. An average of 50–75 million flies were produced per week for the duration of the eradication programme and 400 flies per box, dispersed at the rate of 76–304 sterile flies/km^2 from single-engine planes flying at a height of about 500 m, treating 3.2 km-wide swaths in a grid pattern (Smith, 1960). Some introductions of animals from South America on which the fly could be detected have been followed up by localized release of the sterile flies. The technique was also used when the fly was detected in Libya to prevent it spreading to wild animals in sub-Saharan Africa. Other successes with SIT have been achieved for control of fruit fly pests, most particularly the Mediterranean fruit fly (*Ceratitis capitata*) and the Mexican fruit fly (*Anastrepha ludens*).

Adverse effects caused by irradiation needed to sterilize some insects means that they would not be competitive with the wild populations, so other techniques are under development. The first new technique with mosquitoes has been with the release of three to six day-old male adults of the self-limiting strain OX513A of the *Aedes aegypti* with a dominant lethal gene (RIDL) (Alphey and Andreasen, 2002) three times a week to mate with wild females. Their offspring die before becoming adults, so with repeated releases of sufficient numbers of these self-limiting males, there is a reduction in the wild population interrupting disease transmission. Efficacy trials in the Cayman Islands and Brazil have demonstrated that release of the self-limiting mosquitoes can reduce the wild population by more than 90%. The releases, however, must be repeated to maintain control of the wild mosquitoes that reinvade the treated area. More recent research has indicated that if only the daughters of released males die, the surviving sons can affect the mosquitoes previously un-affected by releases. The technique of multiple releases is equivalent to releases of predators or parasitoids regularly on crops, especially those within glasshouses. Further studies using the RIDL method have been with the diamondback moth, so releases will be tried on a larger scale.

Other studies are examining different approaches to modification of mosquitoes depending on whether the effect is to reduce the number of female mosquitoes, reduce their lifespan or prevent the malarial parasite from developing to reduce disease transmission (Burt, 2014). Similar to the RIDL technique, repeated releases are necessary, so to keep the number required for an initial release as low as possible, it may be necessary to apply a series of space treatments of an insecticide with no residual activity to reduce the wild population to a low level, prior to inundative releases to increase the prospect of a highly significant reduction in the wild population. Other research is seeking development of a self-sustaining system needing only a few releases of relatively fewer mosquitoes.

Biological Control

Classic

The sending of 129 (some reports say 514) ladybirds – vedalia beetle (*Rodolia cardinalis*) – from Australia in 1888 to control cottony cushion scale (*Icerya purchasi*) on citrus in California is a classic example of a predator that could survive and suppress a pest population within a year in a different country. The effective control can easily be disturbed by later applying an inappropriate insecticide, such as DDT, as the scale numbers soon increased when DDT was applied 70 years later in the area and reduced the numbers of vedalia beetles (de Bach, 1974). Inoculative biological control has been achieved against many other pests, but notably where the pest is on a geographic island or in a region that is ecologically separate from other farming areas. There have been some examples of success on extensive areas, such as the release of the neotropical parasitoid *Apoanagyrus (Epidinocarsis) lopezi* to control the cassava mealybug *Phenacoccus manihoti*, which had been inadvertently introduced with cassava plants from South America. The releases of *E. lopezi* in 26 African countries reduced the population density of *P. manihoti* sufficiently in most farmers' fields. Neuenschwander (2001) discussed the cost of the extensive search in South America for a suitable parasitoid and the subsequent quarantine measures needed before it could be released. Great care is taken now to avoid the situation when a biological agent, such as the cane toad *Rhinella marina* from Hawaii, was released in Australia to control the native grey-backed cane beetle *Dermalepida albahirtum* and subsequently became a serious pest.

Inundative

In Uzbekistan, concern over the health of large sectors of the population after aerial sprays of toxic insecticides and defoliants had been applied to cotton fields, led to a demand for biological control. The Soviet government set up biofactories to produce *Trichogramma* by rearing the parasitoid on *Sitotroga cerealella* cultured on grain. The aim was to release the *Trichogramma* when bollworms were detected in the cotton fields (Figs 8.17–8.20). At the time, the 1980s, cotton was the main crop alongside areas of lucerne as a fodder crop, which was an important source of

Fig. 8.17. Rearing *Sitatroga cerealella* in Uzbekistan for culturing *Trichogramma.*

Fig. 8.18. Rearing parasitoids in Uzbekistan.

natural enemies to migrate into the cotton. Other crops occupied only very small areas. Bollworm infestations were generally low due to extremely cold winters as well as the presence of natural and released parasitoids. The effectiveness of releases was doubtful as the parasitoid was attuned to parasitizing the cereal moth eggs. There was also some production of the larval parasitoid *Bracon.* After independence, the Uzbeks wanted to grow maize, wheat and other crops. Pesticide use has increased again, although different chemicals are now used.

Biological control over the last few decades has been mainly in protected cropping and has been achieved by a series of releases of the predator or parasitoid. Good examples have been the release of predatory mites (*Phytoseiulus persimilis*) to control spider mites on crops such

Fig. 8.19. Releasing *Trichogramma* in Uzbekistan.

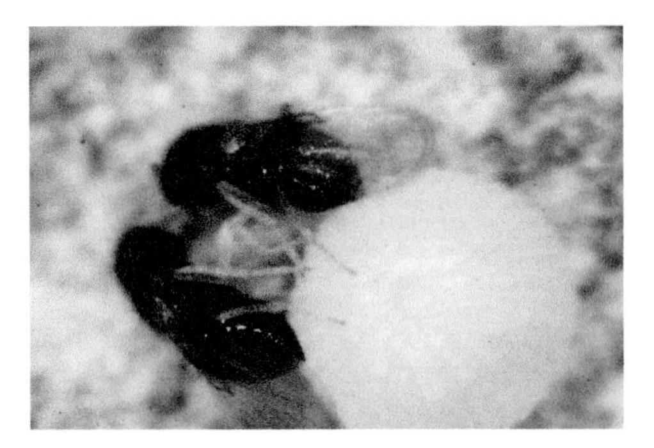

Fig. 8.20. *Trichogramma* wasps on a *Helicoverpa* egg.

as cucumbers, and release of *Encarsia formosa* to control whiteflies (*Trialeurodes* spp.). Success is often dependent on repeated releases. Considerable development of using biological control has enabled specialized methods of release.

Fig. 8.21a. 'Blister pack' containing pupae of the predatory midge *Aphidoletes aphidimyza*. (Photo courtesy of Bioline AgroSciences)

Fig. 8.21b. Gemini sachet containing *Typhlodromips* (or *Amblyseius*) *montdorensis*. (Photo courtesy of Bioline AgroSciences)

A 'blister pack' contains pupae of the predatory midge *Aphidoletes aphidimyza* (Fig.8.21a). These are harvested and packed into a thin layer of moistened vermiculite. The adult midges emerge and mate within the blister, then escape into the crop through a flap at the back of the blister. Each blister would be hung in the crop in an area infested with aphids. The adults lay eggs amongst the aphids, and these eggs hatch into orange, maggot-like larvae that feed by injecting venom into the aphid's leg joints, then consume the aphid once it is subdued. *Aphidoletes* will attack many different species of aphid.

The Gemini sachet, containing *Typhlodromips* (or *Amblyseius*) *montdorensis*, is a patented product containing two independent cells. Each cell contains a breeding colony of a predatory mite together with another mite to act as prey. Each cell also contains a small hole on the inner face, allowing predatory mites to emerge. The specific advantage of the Gemini sachet is that it is tolerant of overhead irrigation or rainfall – the water runs off the outer surface without contacting the emergence hole. *T. montdorensis* is Australian in origin and is licensed for release in several European countries for control of thrips and whitefly (Fig. 8.21b).

Conservation of Natural Enemies

In the UK, the summer of 1976 was extremely hot and there were vast numbers of aphids attacking wheat in June, but little sign of any natural enemies to control them. Some thought growing the variety Maris Huntsman was attracting more aphids, but it may have been the previous winter conditions or other factors early in the season that had reduced predators, such as the seven-spotted ladybird (*Coccinella 7-punctata*). However, with so many aphids on wheat, the ladybird population increased very rapidly, but not early enough to prevent crop damage. As the wheat fields became less attractive, the ladybird spread everywhere in search of food. Farmers did not leave 'tramlines' through their crops so could not use tractor sprayers without damaging the crop. Frantically they tried to get aerial sprays applied, but these were constrained by insufficient aircraft, so sprays were often applied late. Adoption of tramlines came soon afterwards, following experiments in Germany devising methods of applying fertilizers at different crop stages, rather than before or at sowing. This led to an intensification of cereals with wheat sown in the autumn accompanied by fewer crop rotations and increased nitrogenous fertilizer with increased weed and disease problems, leading to greater pesticide use. This example illustrates the difficulty of relying on biological control by inoculation or inundation methods when vast areas need treatment at the same time.

Nevertheless, there is greater interest in conserving natural enemies within farms by setting aside areas of land with wild plants, which also benefit pollinators (Figs 8.22–8.23). This will be encouraged by subsidies to farmers within the EU who are expected to favour environmental management of land by increasing areas sown with wild flowers. In some areas, rather than considering their own farm in isolation, farmers are collaborating in landscape scale improvements to improve conservation of natural enemies and pollinators. In some countries, farmers have often had a forage crop, such as lucerne (Uzbekistan) and berseem clover (Egypt), which provided an area where natural enemies could increase and migrate into crops such as cotton. In the Philippines, studies showed the importance of conservation of predators in rice ecosystems with a series

Fig. 8.22. Wild flowers sown along a field edge for pollinators.

Fig. 8.23. Field edge with hedge providing a location for natural enemies.

Fig. 8.24. Cages in rice field at IRRI in the 1970s to evaluate effect of predators.

of experiments where natural enemies were encouraged or excluded (Fig. 8.24). In Australia, sprays of an attractant were applied in cotton to encourage natural enemies from strips of lucerne to the cotton fields (Mensah, 1999; Gurr *et al.*, 2004).

Farmers can also use selective insecticides or biopesticides, where these are available, in low doses. Improved timing and application target spray deposits more effectively. One idea was to have strips through fields called 'beetle banks' as midfield refuges for naturally occurring predators through fields. Although polyphagous predators did reduce aphid populations in winter, wheat up to 83 m from the beetle bank did not prevent an aphid outbreak in late summer from increasing rapidly (Collins *et al.*, 2002). In agricultural areas, increasing crop rotation may well be an asset to allow greater survival of natural enemies. More consideration is being given to assessing the effect of insecticides on non-target species (Vasileiadis *et al.*, 2017), and in studies on resistance attention should be given to the effect of sub-lethal doses on other insects in the associated ecosystem (Guedes *et al.*, 2017).

Biopesticides

Now that IPM is in the EU policy for pesticides it is expected that there will be much more research to develop new biopesticides in addition to those discussed in Chapter 2. Some of the specialist companies that have pioneered new pesticides with development of suitable formulations have now been merged with the larger agrochemical companies or have linkages to extend their product range. Regulatory authorities are now developing their procedures for biological control agents, for example the ASEAN Guidelines (Bateman *et al.*, 2014).

As indicated earlier in relation to *Metarhizium acridum*, the correct formulation played a crucial part in its development, as control of locusts requires rapid treatment over very large areas, hence the need for a ULV formulation. Although the conidia can be stored dry for a long period, it requires just-in-time formulation at the location of treatments. More recently, Fang *et al.* (2014) found that by co-inoculating recombinant strains expressing a mode of action against sodium channels and a hybrid toxin that blocked both potassium and sodium channels, it was possible to improve the efficacy of *M. acridum* against acridids by reducing lethal dose, time to kill and food consumption. The recombinant strains did not cause disease in non-acridids. Another *Metarhizium* biopesticide, Met52 OD (*Metarhizium anisopliae*), has been approved on-label for use in protected ornamentals in the UK.

A new biological fungicide, which is active against soil pathogens, has now been introduced. It contains the T-22 strain of *Trichoderma harzianum*, which has been developed by hybridization of two strains that had good disease resistance. It is claimed to be effective in all types of soil

Table 8.3. Advantages and limitations linked to the use of microbial biopesticides. (Adapted from Pertot *et al.*, 2017.)

Advantages	Limitations
They do not leave residues on harvested crop.	The spray deposits have a low persistency.
They can be applied close to harvest.	If not correctly applied their efficacy can be lower than chemicals.
They do not interfere with fermentation, e.g. grapes in making wine.	Their application needs more care (weather conditions, quality of water etc.).
They are generally less toxic to humans and the environment than many chemical pesticides.	In the case of high insect pest/disease pressure, they can be less effective.
They are renewable.	They are generally more expensive than chemicals.
They are biodegradable.	Their effect is often slower than chemicals.
Their mode of action is complex and can be used in anti-resistance strategies.	They cannot be mixed in a spray tank with other pesticides.
They are a useful tool in organic production and integrated pest management.	Shelf life is shorter than for chemicals, unless stored under specific conditions.
They are safe for workers.	They have a strict expiry date.
There is no (or a short) re-entry time after spraying.	Some of them need to be stored at low temperature.
They are not phytotoxic.	Once open, the box/bag should be resealed hermetically to prevent humidity.
They can be applied with a normal atomizer.	The suspension, once prepared, cannot be stored for a long time.

at a range of temperatures and has, no doubt, been developed because of the EU policy of promoting IPM to reduce chemical inputs. T-22 is one of the few strains of *Trichoderma* with a European registration for sale and authorization for use in ecological agriculture, and acts as a protective shield against pathogens preventing fungal diseases of the root, such as *Fusarium, Sclerotinia, Rhizoctonia, Pythium* and nematodes, which helps improve root growth. The botanical biofungicide based on a plant extract of the *Melaleuca alternifolia* (tea tree oil) has also been recommended as a broad-spectrum fungicide with a preventative and curative mode of action to control several diseases such as powdery mildew. Another new biofungicide has been developed using *Bacillus amyloiquefaciens* strain MBI 600. The foliar spray is marketed for use on salad and fruit crops including lettuce, spinach, grapes, wine grapes, strawberries, onions, carrots and tomatoes. In the USA, the EPA has established that there is no requirement for a maximum permissible level for residues of *Bacillus amyloliquefaciens* strain PTA-4838 used as a fungicide, nematicide or plant growth regulator.

Some of the new products have been marketed specifically to fit IPM programmes and also provide a break with chemical insecticides as part of a resistance management strategy. One product is based on *Burkholderia* spp. strain A396 and is marketed to provide multiple modes of action by affecting the insect's exoskeleton or interfering with moulting, thus being effective against a wide variety of chewing and sucking insects and mites without adverse effects on beneficials. Another new product is based on *Chromobacterium subtsugae*, which stops feeding and reduces reproduction to prevent the development of damaging populations of sucking and chewing insects, flies and mites. It has been used on a wide range of crops, including horticultural crops. Their key advantage in IPM is that they fit very effectively with efforts to maintain natural enemies.

There has been particular interest in the codling moth *Cydia pomonella* on apples and other fruit trees. The granulosis virus (CpGV) was first discovered in Mexico in 1963, with the Mexican isolate, CpGV-M now registered in 34 countries worldwide. However, since 2005, resistance to this strain has been detected (Gebhardt *et al.*, 2014). In early experiments in the UK, medium-volume applications (600 l/ha) using a tractor-powered Commandair mistblower, applying 1.2 × 1013 capsules/ha gave nearly 90% reduction in damage by larvae penetrating deep in fruit, but revealed a lack of persistence of spray deposits (Richards, 1984). In the USA, Arthurs *et al.* (2005) used the baculovirus as a complementary treatment while using pheromones in a mating disruption strategy, but it was not so effective where there are small isolated orchards and heavy pest pressure.

Spodoptera nuclear polyhedrosis virus (NPV) commercial products are available in the USA and Europe: SPOD-X containing *Spodopera exigua* NPV to control insects on vegetable crops and cut flowers in greenhouses, and Spodopterin containing *Spodoptera littoralis* NPV, which is used to protect cotton, corn and tomatoes. Considerable research was done in east

Africa to determine whether an NPV could be applied to control outbreaks of *Spoptera exempta*. Grzywacz *et al.* (2008) reported that SpexNPV was effective when applied at 1×10^{12} occlusion bodies (OB)/ha, if it was applied early, when 1st—3rd instar larvae are present. High levels of mortality were detected 3–10 days post-treatment, indicating that SpexNPV is a potential substitute for chemical insecticides in strategic armyworm management programmes. This example is similar to the use of *Metarhizium acridum* as a mycoinsecticide against locusts, referred to in Chapter 3, as the insecticide does not adversely affect other insects or birds in the environment and disposal of old stock is not costly compared with obsolete stocks of chemical insecticides that occur when the insecticides arrive after the infestation has been controlled or migrated elsewhere.

An interesting sequel to the introduction of a parasitoid into Canada from Scandinavia to control spruce sawfly *Diprion hercyniae* was the introduction of a nuclear polyhedrosis virus, specific for spruce sawfly, which became established, so no further measures were needed to control the sawfly (Bird and Burk, 1961).

Entomopathogenic nematodes (EPNs) have also been released in crops to control certain pests. These nematodes are susceptible to desiccation, so are best applied to moist soil. There have been many efforts to formulate them so that they remain moist during and immediately after application and are close to the pest. The failure of EPNs in one case, despite a high-volume application, was due to the person directing the spray; missing sections of the crop allowed the pest to spread throughout the area.

Use of biopesticides is still in its infancy and much research is still needed to translate potential control agents, as shown under laboratory conditions, into commercial products. While there are many advantages in applying a biopesticide, there are, for example, certain limitations, especially when control is needed to happen quickly.

Beneficial Bacteria

A recent development has been exploring the addition of naturally occurring bacteria, such as rhizobia, that have a beneficial impact on plants, making them more healthy and resistant to pests. Initially, rhizobia has been added as a seed treatment for soybean, but in-furrow and foliar sprays are possible to improve uptake of nutrients. The aim is to reduce the need for pesticide use. A linkage between agrochemical companies and new companies with synthetic biology expertise aims to develop beneficial microbes focusing on nitrogen fixation.

Another development has been the use of the bacterial symbiont *Wolbachia*, transferred from *Drosophila* into the mosquito *Aedes aegypti* where it can block the transmission of dengue and zika viruses. Trials in Australia have shown interesting promise in an urban area, provided sufficient male mosquitoes are released and not counteracted by the immigration of *Wolbachia*-free mosquitoes from surrounding areas (Jiggins, 2017).

Chemical Control

Financially, if farmers want to secure a high yield, the use of a chemical control remains a key component of IPM, but it requires careful choice of pesticide, careful timing of an application based on pest prevalence obtained by regular crop monitoring, and improved application to deliver deposits more precisely. The emphasis has been to choose an insecticide that is to some extent selective, controlling the pest while being far less harmful to its natural enemies (Jansen *et al.*, 2008). A list of selectivity of pesticides with their rating in terms of harm to natural enemies is provided on the IOBC web page.

Thus, IPM is an example of sustainable intensification, defined as 'producing more output from the same area of land while reducing the negative environmental impacts and at the same time increasing contributions to natural capital and the flow of environmental services' (Pretty and Bharucha, 2015). Some aspects of modern farming now look forward to crop inspection using drones, monitoring pests with pheromone traps and using global positioning technology to know exactly where pests need to be controlled.

Recognizing that increased food production in recent years has been due, principally, to improved crop varieties and crop protection with conventional pesticides to ensure a stable crop yield, the emphasis has now been put on sustainability (Lamichhane, 2017). The EU funded a project PURE (Pesticide Use-and-risk Reduction in European farming systems with Integrated Pest Management) and has designed and tested IPM solutions both on research stations and farms (Lescourret, 2017) to build a toolbox of approaches and methods for implementing efficient IPM solutions in the challenging European context. The project involves 310 researchers from 23 institutions (15 research institutes, 2 extension services, 5 industries and 1 research management body) across 10 European countries. The project has involved modelling studies regarding pest evolution and resistance to pesticides, biocontrol methods and agroecological engineering, and has examined new technologies to help the implementation of IPM in the EU (Pertot *et al.*, 2017). A key research area has been to put emphasis on biocontrol by examining agents that can enhance the plant's own defence mechanisms and assist with the development and production of new biocontrol agents.

To achieve the aims will require each country to recognize the importance of diversity in the rotation of crops, with policy frameworks to control the use of pesticides within regions to mitigate selection of resistance, if IPM implementation is going to be successful without adverse effects on the environment. Over the last few decades, there has been less attention to crop rotation due to market forces, and no overall control of the pesticides adopted on an area-wide basis within a region. Noticeably, within the EU, there is no recognition of innovative techniques in genetic engineering that could be immensely important in developing more resistant varieties. In contrast, the Council for Agricultural Science and Technology (CAST) report by Ratcliffe *et al.* (2017) has stated that genetic techniques (such as CRISPR-Cas9, RNAi, marker technology, plant-incorporated protectants, and stacked traits) may fit well into integrated systems. They also emphasized that resistance management plans are essential.

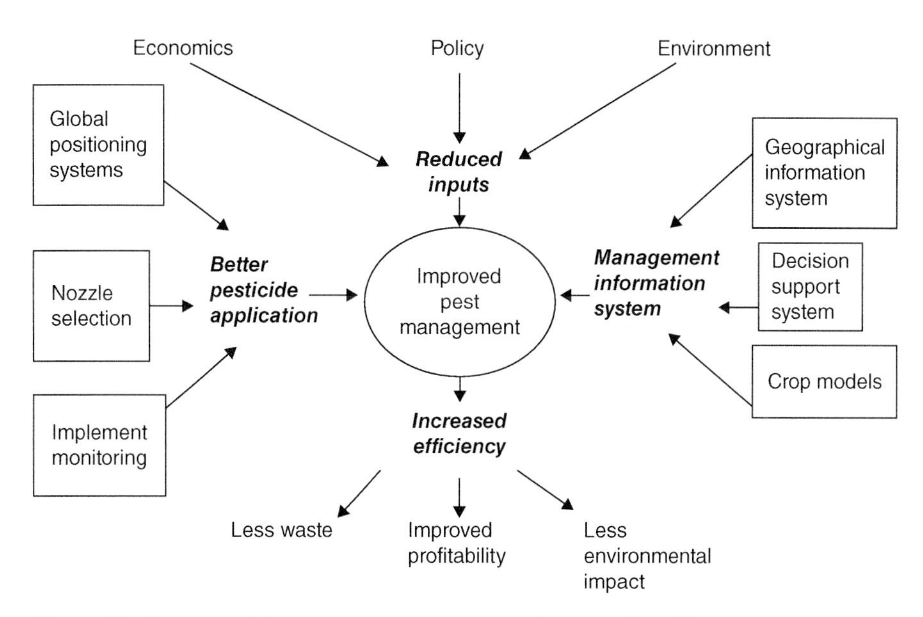

Fig. 8.25. Framework for improvement of pesticide use within IPM.

Despite the objective of endeavouring to avoid using any pesticide, it is likely that some pesticide input will remain crucial for maintaining yields. This is likely to require greater precision in application techniques. Modern technology with sensors to detect changes in the crops will undoubtedly play a key role in minimizing the use of pesticides and directing spot treatments where appropriate.

References

Alphey, L. and Andreasen, M. (2002) Dominant lethality and insect population control. *Molecular and Biochemical Parasitology* 121, 173–178.

Arthurs, S.P., Lacey, L.A. and Fritts Jr, R. (2005) Optimizing use of codling moth granulovirus: effects of application rate and spraying frequency on control of codling moth larvae in Pacific Northwest apple orchards. *Journal of Economic Entomology* 98, 1459–1468.

Bateman, R., Ginting, S., Jan Moltmann, J. and Jäkel, T. (2014) *ASEAN Guidelines on the Regulation, Use, and Trade of Biological Control Agents (BCA)*. German Cooperation.

Bhattacharyya, M. (2017) The push–pull strategy: a new approach to the eco-friendly method of pest management in agriculture. *Journal of Entomology and Zoology* 5, 604–607.

Bird, F.T. and Burk, J.M. (1961) Artificial disseminated virus as a factor controlling the European spruce sawfly *Diprion hercyniae* (Htg) in the absence of introduced parasites. *Canadian Entomologist* 9, 228–238.

Burt, A. (2014) Heritable strategies for controlling insect vectors of disease. *Philosophical Transactions of the Royal Society B - Biological Sciences* 369.

Collins, K.L., Boatman, N.D., Wilcox, A., Holland, J.M. and Chaney, K. (2002) Influence of beetle banks on cereal aphid predation in winter wheat. *Agriculture, Ecosystems & Environment* 93, 337–350.

De Bach, P. (1974) *Biological Control by Natural Enemies.* Cambridge University Press, Cambridge, UK.

Edwards, S. (2015) Push–pull technology in Ethiopia: Progress and Challenges. Available at: https://agriprofocus.com/upload/Push-Pull_Technology_in_Ethiopia_Progress_and_Challenges1448614788.pdf (accessed 24 April 2018).

Eigenbrode, S.D., Birch, A.N.E., Lindzey, S., Meadow, R. and Snyder, W.E. (2016) Review: a mechanistic framework to improve understanding and applications of push–pull systems in pest management. *Journal of Applied Ecology* 53, 202–212.

Fang, W., Lu, H.-L., King, G.F. and St Leger, R.J. (2014) Construction of a hypervirulent and specific mycoinsecticide for locust control. *Scientific Reports* 4, 7345.

Gebhardt, M.M., Eberle, K.E., Radtke, P. and Jehle, J.A. (2014) Baculovirus resistance in codling moth is virus isolate-dependent and the consequence of a mutation in viral gene pe38. *PNAS* 111, 15711–15716.

Grzywacz, D., Mushobozi, W.L., Parnell, M., Jolliffe, F. and Wilson, K. (2008) Evaluation of *Spodoptera exempta* nucleopolyhedrovirus (SpexNPV) for the field control of African armyworm (*Spodoptera exempta*) in Tanzania. *Crop Protection* 27, 17–24.

Guedes, R.N.C., Walse, S.S. and Throne, J.E. (2017) Sublethal exposure, insecticide resistance, and community stress. *Current Opinion in Insect Science* 21, 47–53.

Gurr, G.M., Wratten, S.D. and Altieri, M.A. (eds) (2004) *Ecological Engineering for Pest Management.* CSIRO/CAB International, Wallingford, UK.

Hassanali, A., Herren, H., Khan, Z.R., Pickett, J.A. and Woodcock, C.M. (2008) Integrated pest management: the push–pull approach for controlling insect pests and weeds of cereals, and its potential for other agricultural systems including animal husbandry. *Philosophical Transactions of the Royal Society of London,* B 363, 611–621.

Hoskins, W.M., Borden, A.D. and Michelbacher, A.E. (1939) Recommendations for a more discriminating use of insecticides. *Proceedings of the 6th Pacific Science Congress* 5, 119–123.

Jansen, J.P., Hautier, L., Mabon, N. and Schiffers, B. (2008) Pesticides selectivity list to beneficial arthropods in four field vegetable crops. *Bulletin IOBS/WPRS* 35, 66–77.

Jiggins, F.M. (2017) The spread of *Wolbachia* through mosquito populations. *PLOS Biology* 15(6).

Knipling, E.F. (1979) *The Basic Principles of Insect Population Suppression and Management.* USDA Agriculture Handbook No. 512. United States Department of Agriculture, Washington, D.C.

Kogan, M. (1998) Integrated pest management: historical perspectives and contemporary developments. *Annual Review of Entomology* 1, 243–270.

Lamichhane, J.R. (2017) Pesticide use and risk reduction in European farming systems with IPM: an introduction to the special issue. *Crop Protection* 97, 1–6.

Lescourret, F. (2017) Toward a reduced use of pesticides in European farming systems: an introduction to the PURE project. *Crop Protection* 97, 7–9.

Luck, R.F., Van den Bosch, R. and Garcia, R. (1977) Chemical insect control: a troubled management strategy. *BioScience* 27, 606–611.

Marks, R.J. (1976) The influence of behaviour modifying chemicals on mating success of the red bollworm *Diparopsis castanea* Hrnps. (Lepidoptera, Noctuidae) in Malawi. *Bulletin of Entomological Research* 66, 279–300.

Marks, R.J., Nesbitt, B.F., Hall, D.R. and Lester, R. (1978) Mating disruption of the red bollworm of cotton *Diparopsis castanea* Hampson (Lepidoptera: Noctuidae) by ultra-low-volume spraying with a microencapsulated inhibitor of mating. *Bulletin of Entomological Research* 68, 11–29.

Mensah, R.K. (1999) Habitat diversity: implications for the conservation and use of predatory insects of *Helicoverpa* spp. in cotton systems in Australia. *International Journal of Pest Management* 45, 91–100.

Metcalf, R.L. (1980) Changing role of insecticides in crop protection. *Annual Review of Entomology* 25, 219–256.
Midega, C.A.O., Pittchara, J.O., Pickett, J.A., Hailua, G.W. and Khan, R. (2018) A climate-adapted push–pull system effectively controls fall armyworm, *Spodoptera frugiperda* (J.E. Smith), in maize in East Africa. *Crop Protection* 105, 10–15.
Nesbitt, B.R., Beevor, P.S., Cole, R.A., Lester, R. and Poppi, R.G. (1975) The isolation and identification of the female sex pheromones of the red bollworm moth, *Diparopsis castanea*. *Journal of Insect Physiology* 21, 1091–1096.
Neuenschwander, P. (2001) Biological control of the cassava mealybug in Africa: a review. *Biological Control* 21, 214–229.
Parnell, F.R., King, H.E. and Ruston, D.F. (1949) Insect resistance and hairiness of the cotton plant. *Bulletin of Entomological Research* 39, 539–575.
Pertot, I., Caffi, T., Rossi, V., Mugnai, L., Hoffmann, C. *et al.* (2017) A critical review of plant protection tools for reducing pesticide use on grapevine and new perspectives for the implementation of IPM in viticulture. *Crop Protection* 97, 70–84.
Pimentel, D., Krummel, J., Gallahan, D., Hough, J., Merrill, A. *et al.* (1978) Benefits and costs of pesticide use in U.S. food production. *BioScience* 28, 772–784.
Pretty, J. and Bharucha, Z.P. (2015) Insect pest management for sustainable agriculture in Asia and Africa. *Insects* 6, 152–182.
Ratcliffe, S.T., Baur, M., Beckie, H.J., Giesler, L.J., Leppla, N.C. and Schroeder, J. (2017) Crop protection contributions toward agricultural productivity. CAST Issue Paper 58.
Richards, M. (1984) The use of a granulosis virus for control of codling moth, *Cydia pomonella*: application methods and field practice. PhD thesis, University of London.
Ripper, W.E. (1956) Effect of pesticides on balance of arthropod populations. *Annual Review of Entomology* 1, 403–438.
Ripper, W.E., Greenslade, R.M. and Hartley, G.S (1951) Selective insecticides and biological control. *Journal of Economic Entomology* 44, 3448–3459.
Smith, C.L. (1960) Mass production of screwworm (*Callitroga hominivorax*) for the eradication program in the southeastern states. *Journal of Economic Entomology* 53, 1110–1116.
Stern, V.M., Smith, R.F., van den Bosch, R. and Hagen, K.S. (1959) The integration of chemical and biological control of the spotted alfalfa aphid. I: The integrated control concept. *Hilargdia* 29, 81–101.
Tunstall, J.P., Sweeney, R.C.H. and Matthews, G.A. (1959) Cotton insect pest investigations in the Federation of Rhodesia and Nyasaland. Part I: Cotton bollworm investigations. *Cotton Growing Review* 36, 268–275.
Van Den Bosch, R. and Stern, V.M. (1962) The Integration of chemical and biological control of arthropod pests. *Annual Review of Entomology* 7, 367–386.
Van Emden, H.F. (1972) Plant resistance to insect: developing risk-rating methods. *Span* 15, 71–74.
Vasileiadis, V.P., Veres, A., Loddo, D., Masin, R., Sattin, M. and Furlan, L. (2017) Careful choice of insecticides in integrated pest management strategies against *Ostrinia nubilalis* (Hübner) in maize conserves *Orius* spp. in the field. *Crop Protection* 97, 45–51.
Witzgall, P., Kirsch, P. and Cork, A. (2010) Sex pheromones and their impact on pest management. *Journal of Chemical Ecology* 36, 80–100.

9 Health Issues

The discovery of DDT initiated a careful examination of its mammalian toxicity so that it could be used to protect troops during World War II, especially as many were dying of malaria or were so ill, as if wounded, and thus not able to fight. Samples were obtained from Geigy in Switzerland for experimental use organized in the UK by the Ministry of Supply, which was credited with being the first to use the name DDT in 1943. Other testing was being done at the same time in the USA.The problem of malaria was particularly acute in the Pacific region, where in 1942 about 24,000 of the 75,000 American and Filipino soldiers involved in the campaign to prevent the Japanese advance were suffering from malaria. The military soon learnt that effective malaria control was essential for the successful conclusion of the campaign in the Pacific. There was reliance on quinine to treat those with malaria, as chloroquine had not yet been considered safe to give to humans and the alternative anti-malaria drug was quinacrine, available as Atabrine, but this had serious after-effects – nausea, headaches and diarrhoea. In 1943, the army began using DDT as a 5% dust applied directly to soldiers and refugees in Italy to combat a typhus epidemic as this treatment was highly effective against the lice that carried the disease. The military soon realized that DDT could also be useful against malaria. Later, as the war ended, campaigns against mosquito vectors of malaria expanded in Italy and elsewhere.

Post-war, Hayes (1959) documented information on the pharmacology and toxicology of DDT. Among the 685 references in this publication there are two interesting reports of pure DDT being given to a person. 'One man weighing 74 kg took 250 mg of pure DDT 3 times a day for 3 days without noting any effect on his well-being.' Another case reported 1500 mg of DDT in butter was eaten without effect and lice were killed when they were experimentally fed on his body six and twelve hours after the dose was taken, but not when fed 36 hours after the dose was given. Not surprisingly, DDT was regarded as safe to use and became the key weapon against the mosquito vectors of malaria. Its downfall was the persistence of the chemical in the environment and accumulation down the food chain affecting birds and other non-target organisms. A more recent assessment of human health aspects of using DDT (WHO, 2011) was published, as DDT has been used again in indoor residual spraying.

As mentioned earlier, following the publication of *Silent Spring*, the ban on persistent organic pesticides, notably the organochlorine insecticides, led to greater use of organophosphate and other insecticides, many of which were much more toxic to humans. The arrival in the UK of demeton-S-methyl was a particular concern as it was applied to Brussels sprouts. The reports of the Zuckerman Committee, discussed in the next chapter, focused on parathion (WHO class Ia) and the herbicide DNOC (WHO class Ib) and made strong recommendations on personal protective clothing. This was due to two fatalities arising from the use of DNOC that were reported in 1950, as both men, employed by contractors, had failed to wear the protective clothing (rubber boots, rubber gloves, rubber apron and eye shield) with which they had been provided. They had also worked excessively long hours in hot weather.

In the UK, the Pesticide Safety Precaution Scheme had a significant impact on what pesticides were approved for use in the UK. An independent Pesticide Incident Appraisal Panel (PIAP) was set up to investigate reports of incidents of alleged ill health resulting from the use of pesticides as part of the post-approval surveillance process. There is also a Wildlife Incident Investigation Scheme (WIIS) to identify potential misuse of pesticides affecting wildlife. In many countries baits with pesticides have been used to kill predators and other animals that affect livelihoods, including domestic and wild animals.

The Importance of Application Practices

Training of those employed in the pesticide industry was initiated in the UK with a scheme known as BASIS, in 1978. Those applying pesticides, except domestic and garden products, must have a certificate of competence in using pesticide application equipment. This has led to the setting up of the National Register of Sprayer Operators (NRoSo) in the UK, whereas in many countries there has been no coordinated effort to provide practical training and examination of the operator's ability to follow the requirements for safe application of pesticides. In consequence, more people have reported being ill in the least developed countries, often because of access to highly hazardous pesticides (Loevinsohn, 1987; FAO, 2016), lack of training and failure to use protective clothing. While labels meet requirements for information about first aid and now have pictograms to help convey instructions, there is seldom advice on how the product should be used and what personal protective equipment (PPE) should be worn. Nevertheless, there are claims of chronic renal failure in agricultural communities in parts of

Asia, while other studies have questioned whether the glyphosate and the adjuvants, or the combination of the two, are the basis of the observed kidney and, especially, liver toxicity, as adjuvants are not subjected to the same registration approval process (Mesnage and Antoniou, 2018).

There are many scientific papers that refer to various illnesses that are the result of exposure to pesticides, but many rely on interviews and seldom provide evidence of the extent of exposure. Mostafalou and Abdollahi (2017) examined 14,390 published reports and selected 7419 that met certain criteria. These were that the publication was in English, the study was a cross-sectional, case-control, cohort, ecological and/or a meta-analysis, with exposure assessment based on interviews, questionnaires, geographic information systems (GIS), job exposure and/or a residue detection in biological samples. The last factor was a reported association of chronic diseases due to exposure to pesticides. There is no indication in their assessments about the extent to which individuals were exposed to a pesticide, or the circumstances in which they were exposed. They demonstrated that most of the associations between pesticides and one of the 43 human diseases considered was due to exposure to insecticides, and most were considered to be due to carcinogenicity, followed by metabolic toxicity, reproductive toxicity, developmental toxicity, pulmonotoxicity and neurotoxicity. Apart from insecticides, there has been controversy about the herbicide glyphosate, which was considered to be probably carcinogenic to humans by the International Agency for Research on Cancer (IARC). This relates to conflicting evidence on whether glyphosate leads to non-Hodgkin lymphoma, a cancer of the lymph nodes, in workers who handled the herbicide. It was concluded that the 'overall weight of evidence indicates' glyphosate is not genotoxic in mammals at doses and routes 'relevant to human dietary exposure'.

Being associations, the disease could easily have been due to factors unrelated to the undefined exposure to pesticides. In practice, the most exposures to humans could occur during manufacture, packaging or in field use, while preparing a spray with the concentrated active substance. Once the pesticide is diluted, it has been argued that the user's exposure is to only a diluted pesticide, but there remain major differences between those who have knapsack sprayers and are far more exposed to the spray when they hold their nozzle in front of their body as they walk through a crop, compared to a tractor driver, usually protected within the cab (Table 9.1). Others can be exposed by walking through a treated field, perhaps while it is sprayed (referred to as a bystander), or by harvesting a treated crop and touching dried residues on plant surfaces, or living alongside areas treated with pesticides (resident). In homes, exposure is also to spray from small aerosol cans that are used to control cockroaches and other insect pests. Unless the true extent of exposure is known, the cause of a disease may be quite different, including exposure to exhaust gases from vehicles, or a hereditary factor.

In contrast to the situation in the UK, parathion, in particular, and other highly hazardous organophosphorus insecticides developed later,

Table 9.1. EFSA guidance on contrast between tractor and knapsack spraying, expressed as mg exposure/kg-active substance applied.

	75th centile		95th centile	
	Dermal exposure			
	Hands	Body	Hands	Body
Tractor boom sprayer	0.730	0.917	10.6	4.71
Knapsack sprayer	611	1777	2856	10,949
	Inhalation exposure			
Tractor boom sprayer		0.0107	0.0781	
Knapsack sprayer		0.783	5.99	

continued to be used as sprays in many parts of the world. Clearly, the spray operator should have worn appropriate protective clothing, yet in many countries these pesticides were applied to crops regardless of whether PPE was worn. In a survey in Ethiopia of 600 farmers and farm workers in three farming systems – large-scale greenhouses, large-scale open farms and small-scale irrigated farms – 85% of workers did not have any pesticide-related training. Only 10% had a full set of personal protective equipment. Among those who applied pesticides, most, who had received some training, worked in glasshouses (Negatu *et al.*, 2016).

Apart from poisoning the users, these insecticides also became a means of committing suicide. When these insecticides and subsequently developed products with a similar toxicity were banned in Sri Lanka, the number of suicides declined (Fig. 9.1) and Gunnell *et al.* (2017) report that national bans on highly hazardous pesticides, which are commonly ingested in acts of self-poisoning, seem to be effective in reducing pesticide-specific and overall suicide rates. Efforts were also made to restrict their availability, and in some countries a colour-coded label enabled distributors to sell the most hazardous pesticides only to those who were fully aware of the toxicity of the product. Often pesticides are stored inside a house, mostly to avoid theft, but they need to be locked away to permit access only to the farmer. In the UK, when the Pesticide Safety Precautions Scheme (PSP) was set up in 1957, the most hazardous pesticides were registered only as a granule formulation for soil treatment. With EU legislation, even the most extensively used granule products, such as aldicarb in WHO class Ia, have been withdrawn as being too hazardous in the environment.

Suicides were not always due to extremely toxic insecticides being used. In India, many suicides were by farmers who had not received pesticide application training, used the cheapest spraying equipment and failed to get adequate control of pests in their cotton crops. Advised by local traders, they often replaced an insecticide with a different product (trade name), yet it contained the same active ingredient. The real cause

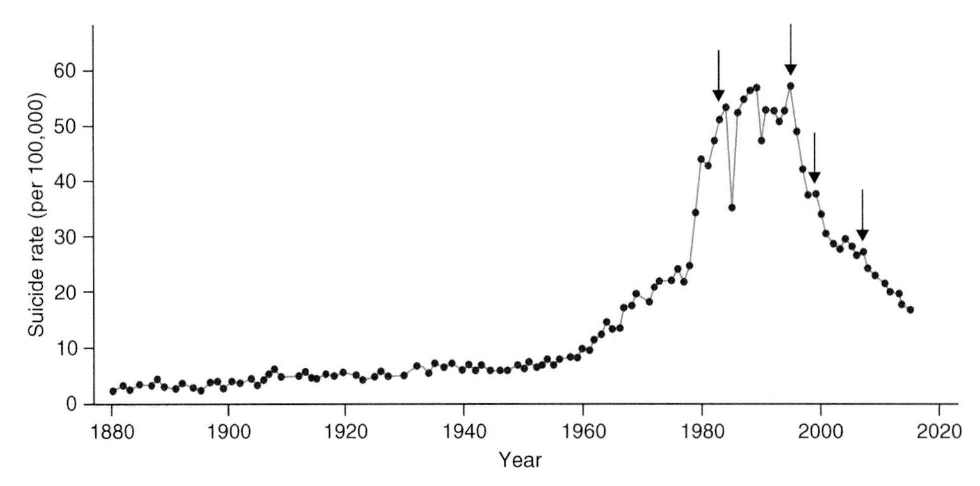

Fig. 9.1. Incidence of suicide in Sri Lanka 1880–2015. Arrows show timing of pesticide bans – parathion, methyl parathion, 1984; all remaining WHO class I toxicity pesticides, 1995; endosulfan, 1998; dimethoate, fenthion and paraquat, 2008. (Redrawn from Knipe *et al.*, 2017.)

of poor control was not necessarily the pest becoming resistant to the pyrethroid insecticides recommended at that time but failure to deposit spray within the crop canopy. The subsequent poor yields resulted in increasing debt and, sadly, to suicide in many cases.

The most serious of incidents that occurred in India was an industrial disaster in 1984, when an extremely toxic gas escaped from the factory manufacturing carbaryl, as mentioned in Chapter 1. This killed nearly 4000 people in the following hours and severely affected many more living in the slums nearby. Sadly, in India, apart from suicides, there have also been many deaths due to applying pesticides, as dealers can be selling all kinds of pesticides that are illegal and not registered for use in the country. In contrast to the BASIS programme in the UK, most of the dealers involved in the illegal trade of pesticides, even if they have a degree in agriculture, do not have the basic knowledge regarding what kind of pesticide should be sold for a specific pest or crop, or how it should be applied.

The problem in many countries is also a lack of applied research to understand the pest situation on crops in a tropical or sub-tropical environment, in contrast to more developed countries in temperate climates, where winter temperatures can impact on pest populations. Where there has been adequate research, clear guidance is possible to determine which pesticides can be applied safely and provide an economic return when applied at the correct time. Sadly, in many countries pesticides continue to be applied without adequate protective clothing and sufficient knowledge about how to apply them safely. Dangers of poor application were pointed out on many occasions, but priority was given, for example in India, to concerns relating to resistance that was building up in the pests (Kranthi *et al.*, 2002) and not about training and improving application techniques.

Fig. 9.2. Spraying cotton near Guntur, India, *c.*1980, with a syringe sprayer, resulting in poor coverage of the crop.

Resistance management in areas showed some reduction in insecticide use, but generally, adoption of IPM was low (Peshin *et al.*, 2009).

The most recent problem, in 2017, occurred on GM cotton with reports from the state of Maharashtra, where at least 50 farmers have died. Although growing Bt cotton was expected to reduce insecticide use, the farmers failed to get good control of bollworms, so sprayed insecticides above the crop while walking into the spray. Subsequent reports suggested that farmers had been using highly hazardous insecticides, including monocrotophos and profenofos. Newspapers also reported that a new Pesticides Bill to regulate the manufacture, quality, import, export and sale of pesticides, to replace the 1968 Insecticide Act, had been pending before the Indian Parliament since 2008.

Spraying a Crop

Small-scale farmers in tropical countries too often have used manually operated knapsack sprayers with the spray lance often fitted with a variable cone nozzle, held in front of the body (Figs 9.3, 9.4 and 9.6) so that as they walk into the spray they are exposed to the spray deposits on foliage. In some countries, especially on rice crops, the nozzles are waved from side to side, so spray aimed upwind is blown back onto the operator when he should hold the nozzle always downwind (Fig. 9.10a and b). Even in the 21st century, some pesticide applications are still made by improvising an application method using perforated containers or a bunch of leaves instead

Fig. 9.3. Spray operator without adequate protection in India, *c.*1970s. The towel around the neck is often used to clean hands.

Fig. 9.4. Spraying rice in the Philippines, a lance being waved from side to side.

Fig. 9.5. Spraying cotton in Egypt with a team pulling the hose through the crop. The helpers carrying the hose are exposed to the spray.

Fig. 9.6. Farmer spraying cotton, incorrectly with lance held at head height and in front of the operator, in the Yavatmal area of India in 2017. (Photo courtesy of Indian Express).

of spray guns or knapsack sprayers, which seem to be too expensive or too heavy, especially for women (Mrema *et al.*, 2017). Women are also likely to be exposed to pesticides if employed to harvest crops, and when washing pesticide-contaminated clothes if they use an empty container that has not been triple-washed.

Fig. 9.7. At shops in Guntur, India, a farmer will see the same active ingredient available with different trade names.

Fig. 9.8. A farmer going to a pesticide shop in Cameroon can get a whole range of products with very similar looking labels, but has he been told which are the least hazardous to use?

Fig. 9.9. Pesticides being sold in Sabah, Malaysia. Note, Paraquat available in a general store.

Fig. 9.10a. Spraying cotton in Pakistan, with lance downwind of operator.

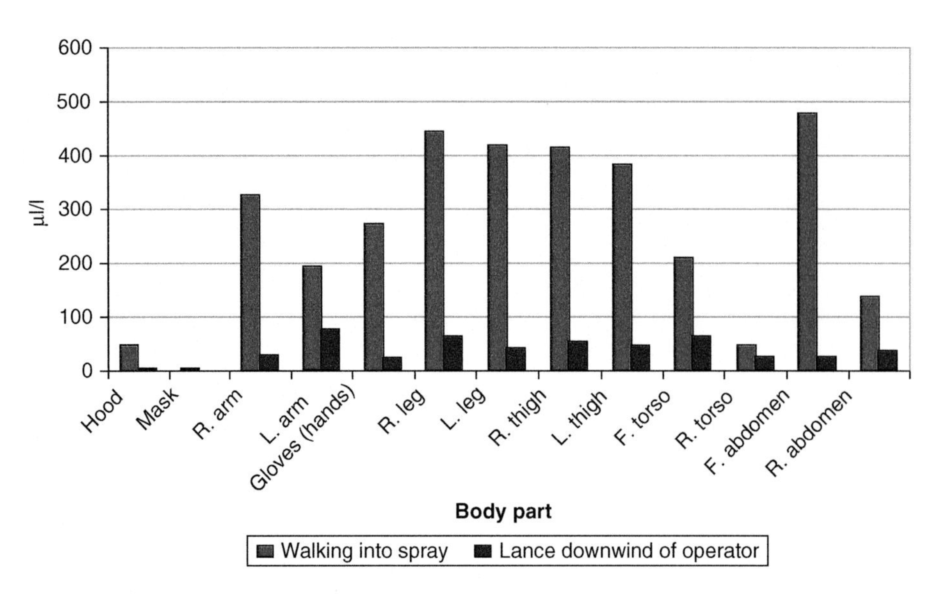

Fig. 9.10b. The impact on exposure by holding the lance downwind rather than straight in front of the body.

When introducing insecticides to cotton farmers in Malawi in 1960, it was decided to mount nozzles on a vertical boom behind the operator, who would walk between rows of unsprayed foliage. Tests (Tunstall and Matthews, 1965) later confirmed that with the tailboom, the main area exposed to spray was around the ankles, which could be easily protected by wearing boots. In contrast, when holding a lance with two nozzles directed laterally into the foliage, more spray was collected on the hands and wrist holding the lance and on the front of the body. When very-low-volume spraying was introduced in west Africa, tests again confirmed less exposure if the spray nozzle was held downwind of the operator rather than in front of the body (Table 9.2).

Mounting the nozzle to the back of the sprayer was successful in central Africa in the 1950s but was never adopted more widely; the weight of the sprayer and extra cost were considered unacceptable, but the importance of protecting the person doing the spraying was not considered. Changing to ultra-low-volume spraying, discussed in an earlier chapter, was more acceptable as obtaining water has remained a problem in arid parts of Africa; but by holding the nozzle always downwind of the body the spray operator was less exposed to the spray.

The ULV technique opened up the possibility of ready-to-use (RTU) pesticide products, but led to large 200 l drums being supplied, in the expectation that local traders could refill the small containers on the sprayer. Often this was not well organized (Fig. 9.11), even though the parastatal companies in francophone west Africa adopted ULV and then VLV spraying of cotton from 1975 onwards.

With conventional spraying using small equipment in Europe, containers of small quantities of pesticide were supplied with a small integral measure so that the quantity was easily measured for a single tank load, without the

Table 9.2. Potential operator exposure with a lever-operated knapsack sprayer in cotton: (a) holding the lance downwind; (b) in front of the operator; (c) a spinning disc VLV sprayer (ULVA+) holding lance downwind, expressed as μl/l of spray applied as determined from fluorescent tracer study. (From Thornhill *et al.*, 1996)

Part of body	Area of disposable overall cm^2	Deposit (a) μl/l	Deposit (b) μl/l	Deposit (c) μl/l
Hood (head)	1200	1.8	45.6	9.3
Mask	172	0.7	3.2	0.05
Right arm	1350	29.7	322.5	63.1
Left arm	1350	76.3	191.0	133.0
Gloves (hand)	900	23.6	269.4	33.6
Right leg	1250	62.7	444.3	11.9
Left leg	1250	42.6	416.2	21.3
Right thigh	1900	52.6	413.3	13.1
Left thigh	1900	45.9	383.2	6.1
Front torso	2750	60.9	209.3	33.9
Rear torso	2750	26.2	45.7	30.4
Front abdomen	3550	25.0	477.4	39.7
Rear abdomen	3550	38.0	139.7	65.8
Total	23,872	486.0	3360.8	461.25

Fig. 9.11. ULV spray in 200 l drums in Cameroon, which had to be transferred to litre containers.

need to pour the liquid into a spoon/measure, which inevitably led to spilling some on the fingers (Figs 9.12, 9.13). This technology was not so readily adopted in developing countries, but some products are now available in sachets. The first sachets were provided in the 1960s with DDT or carbaryl for cotton spraying, but had to be opened to dispense the wettable powder into the sprayer or a separate mixing container (Fig.9.14). Subsequently, a plastic sachet was developed that would dissolve in water so that it did not have to be opened but had to be protected until use inside an outer foil cover. Early versions of the soluble sachets did not dissolve very effectively, especially in

Fig. 9.12. Contamination of fingers while measuring out dose from a bottle using a fertilizer cup.

Fig. 9.13. Exposure of fingers when using a cap from the pesticide container.

Fig. 9.14. Container with built-in measure.

cold water, so threads of plastic blocked filters, but the technology improved. The main purpose of sachets was to ensure the concentration of spray was correct, but it generally reduced operator exposure to the concentrated chemical during preparation of a spray. It also prevented the risk of pesticides being spilt from larger containers in farmers' houses, where their pesticides were often stored to prevent theft.

In many crops, notably rice in south-east Asia, the operator waves the lance from side to side to treat a wide swath, so apart from walking into the spray, he is also exposed to spray drifting downwind when the nozzle is briefly pointing into the wind (Bateman, 2016). Very little is known about the true extent of operator exposure, but instances of leaks from the trigger valve on the lance are reported as well as spillage on the hands during mixing. Operators are also exposed to spillage from the spray tank if the lid is not designed properly with a valve to allow air to enter but liquid to not splash out. Some companies did examine operator exposure for specific pesticides in commercial usage.

Sprayer manufacturers have taken note of health and safety factors, and equipment to meet international standards is more generally available, but the improved equipment tends to be more expensive, so cheaper equipment is still widely used. The only major change in recent years has been to replace the lever-operated knapsack sprayer with a motorized unit, sprayers having an electric motor to drive the pump. This is popular as it cuts the effort needed with manual pumping.

When farmers are spraying their crops, there are often other members of the family, including children, participating in the spraying, as well as others who happen to be walking near the sprayed field (Fig. 9.15). In

Fig. 9.15. Children around a water pump in a village.

many countries there is legislation that prohibits children under 18 years old from being involved with purchasing or using pesticides, but in reality this does not stop children being exposed either in the home to rat poisons or insecticides aimed at cockroaches and other pests, or if they walk in areas with crops which have been sprayed with pesticides. Clearly, the problem is worse in the least developed countries where many children are helping on farms and highly hazardous pesticides are still available.

Residents and Bystanders

Bystanders

A person who happens to be within or adjacent to an area where pesticides are being applied or have just been applied, but whose presence is quite incidental and unrelated to the application of the pesticide, is regarded as a bystander. Originally, it was considered that such a person would be at least 8 m from the spray, as anyone nearer would be part of the spray team and wearing personal protective clothing. More recently, data on spray drift at 2 m downwind of the spray boom have been considered (Fig. 9.16) (Butler-Ellis *et al.*, 2010), as there are public pathways through some fields. Data now shows that bystanders could be exposed to significantly higher amounts of spray than previously considered, but it depends on the nozzles and pressure of the spray being used, as well as the width of the spray boom and the wind conditions. An increase in the use of air induction nozzles emitting a much smaller proportion of droplets below

Fig. 9.16. Measuring spray drift downwind from a tractor sprayer in the UK. Bystanders are at the edge of the field with disposable overalls to measure dye collected on different parts of the body. (Photo courtesy of Silsoe Research.)

Fig. 9.17. Spraying alongside a house in the UK.

100 μm reduces spray drift. Bystanders are at less risk as the smallest droplets remain airborne longer, drift in the wind and may flow around a human body. In the UK, as a precautionary measure, the risk assessment assumes bystanders are exposed to the same daily level of exposure for three months, although sprays would not be applied in the same area so frequently to justify such an approach.

Residents

Concern has been raised by people living in houses close to farmland who have alleged ill health due to pesticides used on neighbouring fields drifting onto their property (Fig. 9.17). Spray drift should not be a problem as the small droplets remaining airborne are unlikely to drift into a house, unless there is air movement through the house, although they may enter gardens. The situation for residents is, however, complicated by other factors, as pesticides are now also used in homes and gardens, both in rural and urban areas. Pesticides deposited on clothing or shoes may also be taken into a house. Most indoor treatments are with small ready-to-use aerosol containers to control ants, cockroaches and flies. These are used sporadically and typically for less than 10 minutes in a small area. In most developed countries, only the least hazardous pesticides are registered for use by the public in their own houses and gardens.

In the tropics, where malaria and other diseases are common, the situation is rather different as the World Health Organization recommends sleeping under a net treated with insecticide. The inside walls of many houses are sprayed to prevent transmission of mosquito-borne diseases (Fig. 9.18). Detailed risk analyses, with the insecticides recommended by WHO, even when spraying DDT on walls, have shown that the residents are not at risk from these applications of insecticides.

In some countries, there have been health problems for those living in poor-quality housing, often within large areas of a crop, such as bananas and cashew nuts, which had been sprayed. To take one example, on a large estate of cashew trees, from 1978 on, endosulfan was sprayed three times a year aerially, using helicopters and small planes. Subsequently, an unusually high number of cancer deaths, neurological disorders and a range of physical and mental impairments were reported (Jayadevan *et al.*, 2005). In 2011, the Supreme Court of India passed an interim order banning the production, distribution and use of endosulfan, as a major number of victims were reported to be affected in Kasargode (Kerala).

Apart from insecticides, certain fungicides, often applied weekly using aircraft, have resulted in claims of ill health. A study in Costa Rica examined the levels of ethylene thiourea (ETU), the main metabolite of mancozeb, on pregnant women near large-scale banana plantations and compared their estimated daily intake with established reference doses. Data from 455 women visited three times during their pregnancy to obtain urine samples showed that median ETU concentrations were more than five times higher than in the general population.

Fig. 9.18. Vector control in India using a stirrup pump and a second person with a lance inside the house to apply an indoor residual spray.

In the UK, a number of families alleged a link between using benomyl while pregnant and their children being born without eyes (anophthalmia) or with related syndromes, including blindness, due to severe damage of the optic stem. In the UK, sachets containing a small quantity of benomyl to mix with water in a garden sprayer had been used on roses, strawberries and tomatoes, so their exposure had been minimal. A case in the USA in 1996, which claimed exposure to a high dose during the formation of the optic nerve in the foetus, resulted in a $4 million award. Studies in which rats were dosed orally at levels of 62.5 mg/kg and higher showed eye defects can occur at relatively high doses. Later, the manufacturer withdrew the product from the market.

Applying Nematicides

Plant parasitic nematodes were causing major losses in banana plantations in central America, so growers were encouraged to apply a nematicide DBCP (Nemagon). It had been approved in the USA, but only later was it realized how dangerous it was to those who had to apply it in the plantations by injecting the fumigant into the soil. It is estimated that about 11 million kg of Nemagon were used each year in the 1960s and early 1970s in central America. Failing to get legal action in their own countries, more than 16,000 banana plantation workers from Costa Rica, Ecuador, Guatemala, Honduras, Nicaragua and the Philippines filed a class-action lawsuit in Texas for compensation from US fruit and chemical companies for permanent sterility linked to exposure to Nemagon (DBCP). In 1997, Amvac, Dow, Occidental and Shell agreed to pay US$41.5 million in an out-of court settlement that resulted in relatively small payments to affected workers. Subsequently, in 2002, a Nicaraguan judge ordered three US companies – Dow Chemical, Shell Oil and Standard Fruit (Dole Food Co. in the US) – to pay US$490 million in compensation to 583 banana workers injured by Nemagon. Use of Nemagon was banned in the USA in 1979.

Workers

Despite increasing robotic systems, many people are employed to harvest crops and flowers, and come into contact with surfaces treated with pesticides as part of their normal working day. Care is particularly needed to avoid entry into and touching treated crops immediately after a spray application. Entry to touch the foliage should be after the pre-harvest interval or restricted entry interval. Hands and other parts of the body touching treated surfaces are most likely to pick up dry, dislodgeable residues so can be protected by wearing gloves and long-sleeved garments to minimize dermal exposure.

Maximum Residue Level (MRL)

To protect the consumer from digesting too much pesticide that could be in harvested crops, the regulators have always insisted on determining the MRL following good agricultural practice with the recommended dosage applied to the crop. Samples of produce have been routinely sampled in a number of countries by government laboratories and by some companies marketing food to ensure that what they sell does not have a pesticide residue that exceeds the MRL.

Food surveys are carried out in EU member states, the USA and some other countries, particularly for certain pesticides, as agreed by member states. EFSA publishes the guidance on the procedures for testing residues in food according to Regulation (EC) No. 396/2005 (Brancato *et al.*, 2017). The UK programme ensures all the major components of our national diet – bread, potatoes, fruit and vegetables, cereals, milk and related products – are sampled. In the UK, the Expert Committee on Pesticide Residues in Food (PRiF) provides independent advice to the government on the monitoring of pesticide residues in food. The risk-based programme is designed to examine commodities likely to contain residues and does not set out to assess residues in our diet. Some commodities are surveyed every year while others are surveyed less frequently, for example once every three years.

Samples from supermarkets, independent shops and market stalls are bought from across the UK by trained purchasers from a market research company, while some samples, including imported food, are taken from a range of points in the supply chain, such as wholesale markets, potato merchants and processors (crisp and frozen chip factories), retail depots, ports and import points. Reports on these surveys are published quarterly and also an annual summary is published for the different commodities sampled. An example of data from a salad crop is shown in Table 9.3.

The 72 samples referred to in Table 9.3 were tested for up to 347 pesticide residues; 27 samples contained no residues from those sought, while 45 samples contained residues above the reporting level. None

Fig. 9.19. Tomatoes with a heavy deposit of spray.

Table 9.3. Number of samples from different types of lettuce, and source.

Commodity	Number of samples taken between January and December 2015	Source
Cos	1	UK
Gem hearts	1	UK
Iceberg	13	UK
	22	EU
Lettuce	5	UK
Little Gem	10	UK
	9	EU
Romaine	4	UK
	1	EU
Round	6	UK

of the samples contained residues above the MRL. Four samples were labelled as organic. Multiple residues were found thus: 21 samples contained residues of more than one pesticide; 8 samples contained two residues; 3 samples contained three residues; 4 samples contained five residues; 1 sample contained six residues; 3 samples contained seven residues; 1 sample contained nine residues; 1 sample contained ten residues. The laboratory detected residues of 23 different pesticides. Following the Chemicals Regulation Directorate's risk assessment, it was not expected that these residues would affect health.

A recent unexpected problem was discovered when eggs were sampled and the insecticide fipronil was detected (www.bbc.co.uk/news/world-europe-40878381). Eggs with this insecticide, which had been exported from Holland, were discovered in 40 countries, 24 of which were within the EU. Millions of eggs had to be taken from supermarket shelves, although many had already entered the food chain via manufacturers of biscuits and cakes. It is extremely unlikely that anyone would have been affected, as a person would have to eat an extremely large number of the contaminated eggs. The cause of the problem was traced to a cleaning company. While fipronil is used to treat pets for ticks and fleas, its use in cleaning buildings used for rearing and holding birds destined for the food supply industry is forbidden.

Nevertheless, many people are concerned about eating food containing pesticides, but as Ames *et al.* (1990) pointed out, plants have evolved with their own defence systems that respond to infestations of pests and thus contain naturally produced chemicals that are effectively pesticides. Most plants that we eat contain natural pesticides, some of which may also be regarded as carcinogens, but the quantities are generally low and measured in parts per billion (ppb). Thus, when we eat various foods, 99.99% of the 'pesticides' that they contain are all natural, but humans have evolved defences to protect against toxins irrespective of whether they are natural or synthetic. We still eat cabbages, broccoli and other vegetables, herbs and spices, and drink coffee, which are liked because they have specific tastes often due to the natural chemicals they contain. It is a great pity that the general public as well as those who promote organic food fail to understand the chemistry of pesticides and that ill effects are due to consuming too high a 'dose'. After all, paracetamol is a popular drug to take away pain/headaches etc., yet it is readily available with a warning on each packet not to exceed a certain number of tablets within 24 hours (Table 9.4).

In a parallel manner to pesticides, the pharmaceutical industry has developed medicines that are derived from some plants, bacteria and fungi that produce substances aimed at relieving pain and killing other microbes, but the latter were called antibiotics. There is now a concern that microbes have become resistant to existing antibiotics, and this is, to a large extent, for the same reason pests have become resistant to pesticides. Natural selection continues when the pesticide or antibiotic is overused.

Table 9.4. Comparison between selected pesticides and chemicals consumed in food and drink.

Chemical	LD_{50} [mg/kg]	Notes
Methyl parathion	25	Insecticide
Lead arsenate	100	Insecticide
Paraquat	150	Herbicide
Caffeine	192	In coffee
DDT	300	Insecticide
Copper sulfate	472	Fungicide
Paracetamol	1944	Painkiller
Glyphosate	>2000	Herbicide
Ethanol	7060	Alcohol
Monosodium glutamate	16,600	Flavour enhancer

Box 9.1. Fruits and vegetables containing natural pesticides that have been found to be carcinogenic to rats.

According to Ames *et al.* (1990), Americans consume an estimated 5000 to 10,000 different natural pesticides every day, many of which cause cancer when tested in laboratory animals; thus:

Carcinogens detected in experiments with rats are present in the following foods: anise, apple, apricot, banana, basil, broccoli, Brussels sprouts, cabbage, cantaloupe, caraway, carrot, cauliflower, celery, cherries, cinnamon, cloves, cocoa, coffee, collard greens, comfrey herb tea, currants, dill, eggplant, endive, fennel, grapefruit juice, grapes, guava, honey, honeydew melon, horseradish, kale, lentils, lettuce, mango, mushrooms, mustard, nutmeg, orange juice, parsley, parsnip, peach, pear, peas, black pepper, pineapple, plum, potato, radish, raspberries, rosemary, sesame seeds, tarragon, tea, tomato, and turnip. Thus, it is probable that almost every fruit and vegetable in the supermarket contains natural plant pesticides that are rodent (rat) carcinogens. The levels of these rodent carcinogens in the above plants are commonly thousands of times higher than the levels of synthetic pesticides present in food.

Protecting Water

A major problem with pesticide use is the need to prevent the chemicals getting into water supplies. The old techniques of application using high volumes of spray inevitably resulted in spray dripping from foliage to the soil where it was exposed to subsequent rain to be either washed from the surface into the nearest ditch or stream or drained down to the water table. Even with lower spray volumes, some pesticide is wasted on the soil and some is lost as spray drift. Buffer zones with grass or other plants are used to reduce the 'run-off' after rain, to prevent it reaching ditches and streams alongside treated fields (Fig. 9.20). Extensive sampling is carried out to check the amount of pesticide found in water and to ensure that levels do not exceed the legal standards (in the EU, 0.1 μg/l of water for a single pesticide and 0.5 μg/l for all pesticides).

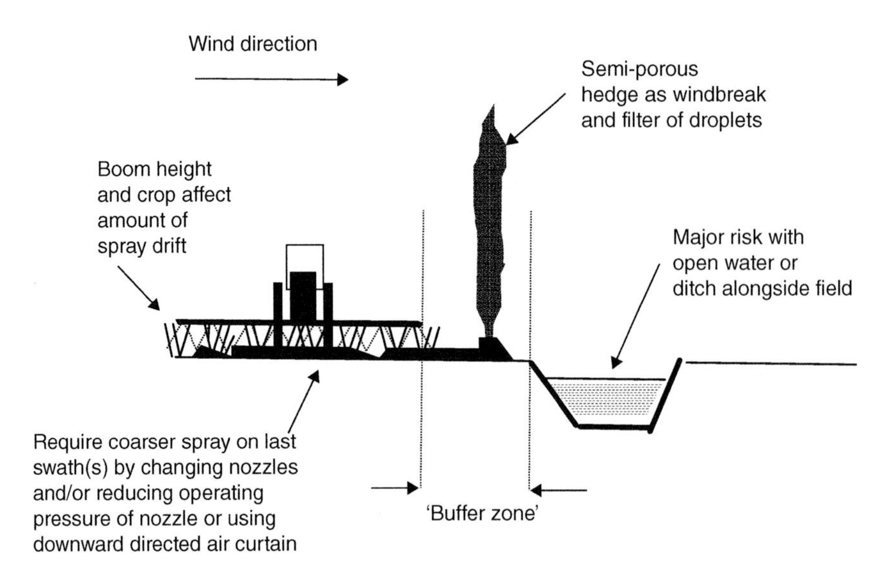

Fig. 9.20. Diagram to show buffer zone to protect ditches and small streams alongside a field.

Washing out sprayers after an application has been a source of pollution, especially when this is done at the farmyard and the waste gets into drains. To reduce pollution, farmers have been advised to wash the sprayer in the field where the pesticide has been applied, facilitated if the sprayer has an extra water tank for cleaning. Another solution has been to create a biobed to collect contaminated water that is degraded by bacteria (Fig. 9.21). There is also water treatment equipment that can be used, particularly by contractors (Fig. 9.22).

In the USA, there has been concern about the Clean Water Act, which established the basic structure for regulating discharges of pollutants into the waters of the USA and regulating quality standards for surface waters. This is because mosquitoes breed in water, so these areas are treated with larvicides. Under the National Pollutant Discharge Elimination System (NPDES) a permit can be requested for any point source discharge of pesticides to waters whether larviciding or adulticiding. The requirement for NPDES has been questioned as it requires additional time to get a permit to use an insecticide already approved by the EPA to be applied over areas with mosquitoes.

Perhaps surprising in a small country, a recent study carried out in Switzerland detected 128 different pesticides (61 herbicides, 45 fungicides and 22 insecticides) across five waterways in catchment areas representative of intense agricultural use. As a result of these findings, the government was asked to reduce the amount of pesticides being used.

The tradition has been to formulate a pesticide in water to facilitate spraying it on crops, but much of the spray reaches the soil, partly at the time of application, but also following heavy rainfall that washes deposits off foliage. Perhaps the agrochemical industry should have adopted

Fig. 9.21. A 'biobed' to collect washings from a sprayer and degrade the pesticides in the water. (Photo courtesy of ADAS.)

Fig. 9.22. A water treatment plant to remove pesticides from water used by sprayers operated by a contractor. (Photo courtesy of Sentinel, used with permission.)

greater use of ultra-low-volume sprays with an oil-based formulation, deposits of which would be less likely to be washed off plants, resulting in less chemical being transferred from fields into waterways.

Fumigation

Gases used in fumigation are known to be highly poisonous, yet there are cases of deaths occurring, especially due to phosphine as it is sold as tablets that generate the gas in contact with moisture. Most cases of poisoning have occurred in tropical countries, but even in the USA there have been cases of deaths due to phosphine poisoning. In one example, in January 2017, ambient humidity released enough phosphine gas from a large quantity of aluminium phosphide tablets under a house to make members of the family feel ill, but later when an attempt was made to wash away the pellets, a large amount of phosphine gas was released, resulting in the death of four children (Balaskovitz, 2017).

Packaging

Some health issues are related to the way a pesticide is packaged. In Chapter 3, the problems of using diazinon as a sheep dip were mentioned, but in tropical countries, where many farmers use manually carried sprayers, the measurement of small quantities for each knapsack load inevitably led to spillage on the user's hands. Apart from the wastage of chemical, the higher concentration of spray was likely to have unacceptable consequences as many farmers do not use protective clothing, not even gloves! While a container with a built-in measure was developed and eventually became standard for the garden market, it was not widely available in the tropics.

Efforts to get pesticide sold in sachets were prolonged as companies had to charge a higher price than if a larger container was used. Having a large packet of a white wettable powder in an African house was considered hazardous as the powder might have been mistaken for flour. In Malawi (then Nyasaland), Union Carbide set up a repackaging system to re-pack Sevin and DDT WP powders, recommended on cotton, into sachets so that one sachet contained the correct amount for mixing with three gallons of water. A local company also made a bucket designed to hold three gallons but fitted with a filter and spout to make it easier to pour the diluted spray into the sprayer tank. This led to the development of a water-soluble packaging that took time to develop as pieces of the packing material 'dissolved', blocking nozzles, especially at low temperatures. In the USA, water-soluble packaging qualifies as a closed mixing/loading system under the Agricultural Worker Protection Standard, when used properly, as it can significantly reduce handler exposure during the mixing and loading of pesticides. However, some unintended practices in the field can actually increase the risks, negating the intention of the technology, so new labelling is required. It is also seen as a means of reducing the amount of packaging that has to be recycled or disposed of as hazardous waste.

Labels now require pictograms to show what protective clothing is required, but these are often quite small on a label (Figs 9.23, 9.24). To provide sufficient information to the farmer a leaflet is often more appropriate (Fig.9.25).

Changes in Formulation

The various types of formulation have been discussed earlier, but a key player in making the changes, industry, was aware of problems when users mixed EC formulations with solvents that could increase dermal uptake of any pesticide spillage on unprotected skin. Similarly, the change from wettable powders to making a wettable granule was to avoid the risk of a puff of dust particles being inhaled by the user.

In labelling products, the manufacturer is keen to advertise their trade name, but many untrained growers fail to see in smaller print what the active ingredients are in the product. This has led to immense problems in some countries, such as India, mentioned earlier in relation to suicides, where there may be several products with the same active ingredients but having different trade names. In some countries, such as China, the active ingredient name must be at least as large as the trade name.

Fig. 9.23. Label with pictograms, Cameroon.

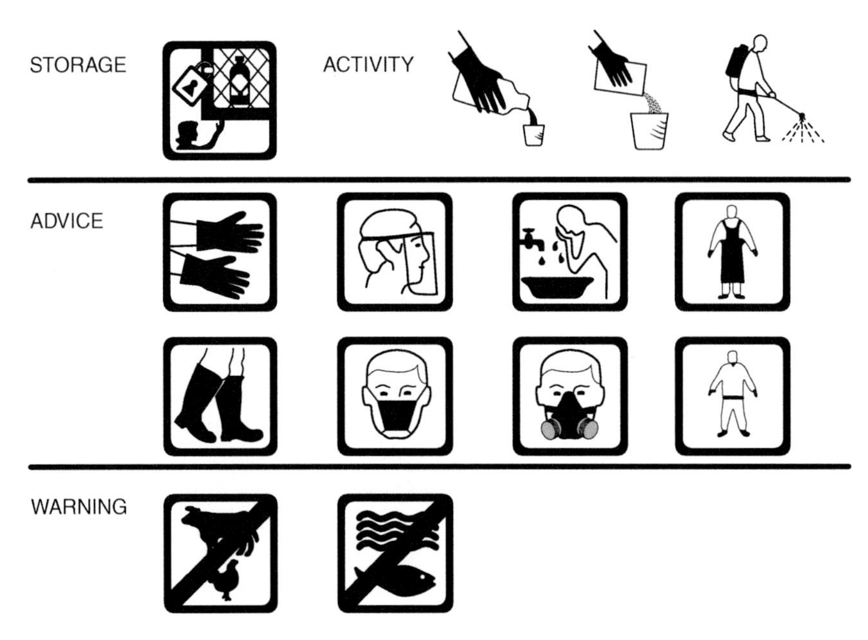

Fig. 9.24. Pictograms. (Photo courtesy of FAO.)

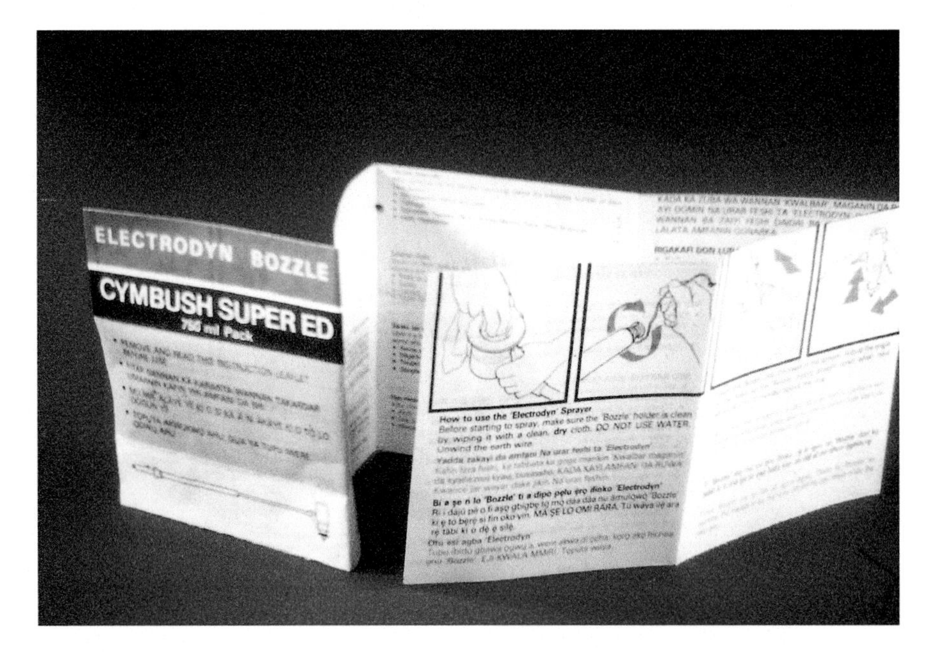

Fig. 9.25. Leaflet to give more information to support the label. (Photo: ICI.)

Misuse of Pesticides

Illegal use of highly toxic pesticides had led to major incidents of poisoning. The largest incidence of pesticide food poisoning documented in the USA was when aldicarb was applied illegally to melons. In 1985, three people who had eaten watermelon in California rapidly became ill with symptoms that included vomiting, diarrhoea, muscle twitches and abnormally slow heart rates. At the same time, people in Oregon were also falling ill. There was an immediate ban on watermelon sales and the crop was destroyed, but a total of 1350 cases of aldicarb poisoning were reported in California, plus another 692 cases elsewhere. Tests confirmed extremely high levels of aldicarb were found in the watermelons, although it was not established if the farmer had used it intentionally or the problem arose from using the pesticide on cotton fields nearby.

Most instances of misuse have occurred when a pesticide is used deliberately to kill wild animals. This may be a rodenticide (see Chapter 6) or an insecticide, such as carbofuran.

Illegal and Counterfeit Pesticide Products

There have been major problems where a country has banned the use of an active ingredient, but it is still used, as irresponsible traders/criminals bring in cross-border supplies illegally. There has also been an increase in poor-quality products with a lower concentration being packaged with a label that is a copy of a recommended product and designed to look like a genuine product. Although the insecticide aldicarb has been banned in South Africa, it has been sold illegally in markets as 'two-step' to quell a rising rat epidemic. Young children have been at risk as the poison is mixed with a cereal to attract the rats to eat it.

Counterfeit pesticides, which are not authorized for sale, may result in the total loss of treated crops, compromising a farmer's livelihood and make the crop unmarketable if unacceptable residue levels are detected. Such products ignore the international labelling requirements designed to ensure safety during transport, as they may contain highly toxic, flammable or otherwise hazardous substances without regard for the safety of people or the environment. Such practices damage the reputation of legitimate stakeholders and challenge sustainable agriculture.

Risk Analysis versus the Precautionary Principle

The EU has adopted a system of registration based on the precautionary principle, yet for over 50 years, recognizing that pesticides are toxic chemicals, they have been used globally based on risk analysis. The most hazardous pesticides (WHO class I), such as aldicarb, were only permitted in the UK if applied as granules with operators wearing protective clothing. As the climate in northern Europe is relatively cool, using PPE was acceptable. Its use in the EU is now banned under the precautionary principle, despite the lack of an alternative less hazardous product of similar efficacy.

In contrast, similarly toxic pesticides that have often been applied in more tropical climates without adequate protective clothing should not have been recommended, as generally users would not be adequately protected. Similar questions are raised about other pesticides, even with much less hazardous use, simply because vocal opposition to pesticides persuades politicians to withdraw registration as a precaution, rather than allow their use with enforcement of safety measures. Recently, issues have been raised about glyphosate and neonicotinoids, as discussed earlier, but in both instances the views of certain groups and the way in which some scientific papers are reported in the media have influenced decisions without a full understanding of the scientific basis on which they were registered as important new products to help farmers improve their crop yields.

Clearly, much has been learnt over the last six decades and the trend is definitely towards using the less toxic pesticides and biopesticides, where effective products are available. In contrast to chemical pesticides, the rigorous regulation of ‘safe’ microbials requires (i) characterization of the isolate, (ii) quality control of the product, and (iii) claims made on the label; whereas botanical extracts should be treated in a similar way to chemicals. However, much more attention is needed on how pesticides are applied, aiming at safer and more targeted systems, and this requires more effective and practical training for both the distributors of pesticides and the farmers.

References

Ames, B.N., Profet, M. and Gould, L.W. (1990) Nature’s chemicals and synthetic chemicals: comparative toxicology. *Proceedings of the National Academy of Science of the USA* 87, 7782–7786.

Balaskovitz, R. (2017) Report says phosphine gas caused death of Balderas kids. Amarillo Globe-News.

Bateman, R. (2016) The role of pesticides in SE Asian rice IPM: a view from the Mekong Delta. *Outlooks on Pest Management* 27, 53–60.

Brancato, A., Brocca, D., Erdos, Z., Ferreira, L., Greco, L. *et al.* (2017) Reporting data on pesticide residues in food and feed according to Regulation (EC) No. 396/2005 (2016 data collection). European Food Safety Authority (EFSA).

Butler-Ellis, M.C., Lane, A.G., O’Sullivan, C.M., Miller, P.C.H. and Glass, C.R. (2010) Bystander exposure to pesticide spray drift: new data for model development and validation. *Biosystems Engineering* 107, 162–168.

FAO (2016) *Guidelines on Highly Hazardous Pesticides.* Food and Agriculture Organization of the United Nations, Rome.

Gunnell, D., Knipe, D., Chang, S.-S., Pearson, M., Konradsen, F., Lee, W.J. and Eddleston, M. (2017) Prevention of suicide with regulations aimed at restricting access to highly hazardous pesticides: a systematic review of the international evidence. *The Lancet Global Health* 5, 1026–1037.

Hayes, W.J. (1959) Pharmacology and toxicology of DDT. In: Simmons, S.W. (ed.) *DDT Vol. II – Human and Veterinary Medicine.* Birkhauser Verlag, Basel und Stuttgart, Germany.

Jayadevan, S., Jayakumary, M., Venugopalan, P.P. and Binoo, D. (2005) Study on health hazards associated with endosulfan spray in cashew plantation in Kerala, India. *Epidemiology* 16, S65.

Knipe, D.W., Gunnell, D. and Eddleston, M. (2017) Preventing deaths from pesticide self-poisoning – lessons from Sri Lanka. *The Lancet Global Health* 5, 651–652.

Kranthi, K.R., Jadhav, D.R., Kranthi, S., Wanjari, R.R., Ali, S.S. and Russell, D.A. (2002) Insecticide resistance in five major insect pests of cotton in India. *Crop Protection* 21, 449–460.

Loevinsohn, M.E. (1987) Insecticide use and increased mortality in rural central Luzon, Philippines. *The Lancet* 1359–1362.

Mesnage, R. and Antoniou, M.N. (2018) Ignoring adjuvant toxicity falsifies the safety profile of commercial pesticides. *Frontiers in Public Health* 5, 361. DOI: 10.3389/fpubh.2017.00361.

Mostafalou, S. and Abdollahi, M. (2017) Pesticides: an update of human exposure and toxicity. *Archives of Toxicology* 91, 549–599.

Mrema, E.J., Ngowi, A.V., Kishinhi, S.S. and Mamuya, S.H. (2017) Pesticide exposure and health problems among female horticulture workers in Tanzania. *Environmental Health Insights* 11, 1–13.

Negatu, B., Kromhaut, H., Mekonnen, Y. and Vermeulen, R. (2016) Use of chemical pesticides in Ethiopia: a cross-sectional comparative study on knowledge attitude and practice of farmers and farm workers in three farming systems. *Annals of Occupational Hygiene* 60, 551–566.

Peshin, R., Dhawan, A.K., Kranthi, K.R. and Singh, K. (2009) Evaluation of the benefits of an insecticide programme in Punjab, India. *International Journal of Pest Management* 55, 207–220.

Thornhill, E.W., Matthews, G.A. and Clayton, J.S. (1996) Potential operator exposure to insecticides: a comparison between knapsack and CDA spinning disc sprayers. *Proceedings of the Brighton Conference*, Vol. 3. BCPC, Farnham, UK, 1175–1180.

Tunstall, J.P. and Matthews, G.A. (1965) Contamination hazards in using knapsack sprayers. *Cotton Growing Review* 42, 193–196.

WHO (2011) *DDT in Indoor Residual Spraying: Human Health Aspects*. IPCS Environmental Health Criteria 241.

10 Regulations and the Manufacturers of Pesticides and Related Organizations

The first country to introduce legislation was the USA when they introduced the Federal Insecticide Act in 1910. The full title was 'An Act for preventing the manufacture, sale, or transportation of adulterated or misbranded Paris greens, lead arsenates, and other insecticides, and also fungicides, and for regulating traffic therein and other purposes'. This Act stood the test of time, but following World War II, the application of synthetic organic insecticides increased from 100 million pounds in 1945 to over 300 million pounds by 1950, so in 1947 Congress passed the Federal Insecticide, Fungicide, and Rodenticide Act to address some of the shortcomings of the previous Act and address the growing issue of potential environmental damage and biological health risks associated with such widespread use of pesticides. Then the responsibility was moved from the Department of Agriculture in 1972, when the Federal Environmental Pesticide Control Act (FEPCA) was enacted and the Environment Protection Agency (EPA) was set up. Subsequently, in 1988, an amendment required re-registration of many pesticides that had been registered before 1984, followed in 1996 by the Food Quality Protection Act, and in 2012 by the Pesticide Registration Improvement Extension Act.

In the UK, in 1950, the government asked Prof. Solly Zuckerman to chair a Working Party on Precautionary Measures against Toxic Chemicals, as there was concern about the risks to users, to food consumers and to wildlife. Up until 1951, seven agricultural workers were known to have died as a result of dinitro-ortho-cresol (DNOC) poisoning and there were concerns about sprays of parathion. The first report from the Working Party published in that year led directly to the Agricultural (Poisonous Substances) Act 1952, which aimed to protect agricultural workers from the most toxic products by requiring that protective clothing be worn when using pesticides and restricting the hours workers were permitted to work with them. The report contained considerable detail on the need for personal protective clothing and hours of operation. It led to the introduction of the Notification Scheme (subsequently called the Pesticides Safety Precautions Scheme (PSPS)) in 1955. The Safety Scheme was a voluntary arrangement agreed between industry and the government departments concerned. About that time, a highly toxic organophosphate insecticide – demeton-S-methyl (Metasystox) – was used to control aphids on Brussels sprouts, and in 1954, following advice from the Working Group, the government established the Advisory Committee on Poisonous Substances Used in Agriculture and Food Storage. The Medicines Act 1968 removed the veterinary medicines, so the committee became the Advisory Committee on Pesticides (ACP) and a principal source of advice on pesticide safety issues. The committee was composed mainly of medical specialists, but later it had two sub-committees to deal with residues in food and environmental issues. Since 2015, the ACP has been renamed the Expert Committee on Pesticides (ECP).

The approval of pesticides continued to be through the Pesticides Safety Precautions Scheme (PSPS), which operated from 1957 for agricultural products, but later became a statutory scheme with the Food and Environmental Protection Act 1985 (FEPA). Non-agricultural products were included in the 1970s as part of the remit of the Health and Safety Executive (HSE). The Pesticide Safety Directorate (PSD), which had been with the Ministry of Agriculture and then the Department of Food and Rural Affairs, became part of the Chemical Regulation Directorate (CRD) set up by the HSE. With changes in legislation within the EU, CRD became the competent authority to regulate chemicals, pesticides, biocides and detergents within the UK.

The effect of setting up the ACP was that insecticides classified by the WHO as extremely hazardous (class I) were no longer approved for application as sprays, although certain chemicals did receive approval applied as granules to the soil. Aldicarb, phorate and disulfoton were approved if the granules contained less than 10% of the active ingredient with no dust present. Meanwhile, in many countries, extremely hazardous chemicals were sprayed, often without any personal protection. Long after parathion was no longer registered in the UK it was used on cotton in the USA and in central America, often mixed with DDT and toxaphene.

Within the EU, harmonization of the approval of pesticides resulted in a member state acting as a *rapporteur* to assess the data of an individual pesticide submitted by the company producing the chemical before a decision is made to approve it. The European Commission Regulation (EU) No. 283/2013 of 1 March 2013 sets out the data requirements for active substances, in accordance with Regulation (EC) No. 11107/2009 of the European Parliament and the Council, which relates to placing plant protection products on the market. Active substances that do not have an unacceptable risk to people or the environment are added to the list of approved active substances contained in the Commission's Implementing Regulation (EU) No. 540/2011. This Regulation follows the European

Commission's use of the precautionary principle to protect human health and the environment. However, this has created considerable debate as it does not assess the risk of exposure to the active substances and adopts a hazard assessment. In consequence, a large number of previously registered pesticides have been withdrawn from use. While the withdrawal of the extremely hazardous pesticides is logical, there are some active substances that, by formulation, such as a granule or seed treatment, can be used effectively with minimal exposure to trained users. Once the active ingredient is approved, member states can register products/formulations that can be marketed within their country. The European Food Safety Agency (EFSA) was set up in February 2002 and is based in Parma, Italy. It covers the direct or indirect impact on food and feed safety, including animal health and welfare, plant protection, and plant health and nutrition, and thus examines the safety of pesticides, especially in relation to residues in crops, while the Health and Safety DG looks at overall risk management.

The harmonization of approvals has led to decisions that are not always acceptable to all member states. Whilst the removal of the registration of many old products, especially those considered to be highly hazardous, has in most cases been welcomed, inevitably there are exceptions where the pesticide has a niche market. Asulam sodium salt, used for bracken control and marketed as Asulox, is a particular example where, at present, there is no other suitable herbicide. Other decisions that have caused considerable controversy relate to the banning of neonicotinoids, as discussed in Chapter 3. This has been based largely on public disquiet and on anti-pesticide lobbying of politicians rather than informed scientific discussion between all parties of a very complex issue. Bee colonies are affected by mites, viruses, food supply and weather conditions, as well as the risk of spray drift exposing flowering plants downwind to droplets that could also be collected by foraging bees. The important factor is that the neonicotinoid insecticide need not be sprayed, but applied as a seed treatment to protect crops during their early stages of growth. Quality and method of seed treatment, and type of equipment employed to sow the treated seed, were factors that led to the original moratorium. Improved seed treatment to minimize any dust, and using seeders that did not distribute dust into the atmosphere, were essential developments since the first use of these important insecticides.

Although one aim of the EU Regulations was increased free movement and availability of plant protection products, there are still differences that hinder equal competition within the Common Market. In a sample examination of the products available in the Czech Republic, Germany, Hungary, Lithuania, Poland and Slovakia in 2016, there are differences in the number of active substances available for oilseed rape and potato protection, with only about half of the products available in all five countries (Matyjaszczyk and Sobczak, 2017).

Each country is responsible for having legislation that controls the use of pesticides within their boundaries, but many of the smaller, less developed countries have insufficient trained people or funds to implement controls on their use, such as the application of pesticides that have been imported illegally and have not been registered. Lack of training of distributors of pesticides, and the many farmers, are also crucial problems, with many farmers applying highly hazardous chemicals without appropriate protection. In 1985, the Food and Agriculture Organization (FAO) decided to publish an International Code of Conduct on the Distribution and Use of Pesticides in support of increased food security, but also aimed at protecting human health and the environment. Subsequently, a number of guidelines were published (Table 10.1), some of which have been revised, and the Code was updated in 2006 and revised in 2014 as the International Code of Conduct on Pesticide Management.

The FAO has, in cooperation with UNEP and others, developed a toolbox to assist those responsible in developing countries for registration to have access to information that is needed in deciding on whether a pesticide should be approved for use. The FAO Pesticide Registration Toolkit is a decision support system provided as a web-based registration handbook to assist registrars in the evaluation and authorization of pesticides.

Registration staff can use the toolkit to support several of their regular tasks, including: finding data requirements; evaluating technical aspects of the registration dossier; choosing appropriate pesticide registration strategy and procedures; reviewing risk mitigation measures; and getting advice on decision-making. The toolkit also links to many pesticide-specific information sources such as registration in other countries, scientific reviews, hazard classifications, labels, MRLs and pesticide properties.

International Conventions

The Basel Convention on the Control of Transboundary Movements of Hazardous Wastes and Their Disposal was adopted in 1989 in response to concerns about toxic waste from industrialized countries being dumped in developing countries and countries with economies in transition. The Convention's principal focus was to elaborate the controls on the 'transboundary' movement of hazardous wastes and the development of criteria for environmentally sound management of wastes. More recently, the work of the Convention has emphasized full implementation of treaty commitments, promotion of the environmentally sound management of hazardous wastes and minimization of hazardous waste generation. The Convention became effective on 5 May 1992.

Rotterdam Convention

One of the concerns about movement of highly hazardous pesticides was that some countries were importing pesticides that were likely to cause

Table 10.1. List of guidance documents published by the FAO.

1. Guidelines for Legislation on the Control of Pesticides (1989)
2. Guidance on Pest and Pesticide Management Policy Development (2010)
3. FAO/WHO Guidelines for the Registration of Pesticides (2010)
4. FAO/WHO Guidelines on Data Requirements for the Registration of Pesticides (2013)
5. Guidelines on Efficacy Evaluation for the Registration of Plant Protection Products (2006)
6. Guidelines on Good Labelling Practice for Pesticides (1995)
7. Guidelines on Environmental Criteria for the Registration of Pesticides (1989)
8. Guidelines on the Registration of Biological Pest Control Agents (1988)
9. FAO/WHO Guidelines for Quality Control of Pesticides (2011)
10. Guidelines on Compliance and Enforcement of a Pesticide Regulatory Programme (2006)
11. FAO/WHO Guidelines on Pesticide Advertising (2010)
12. Provisional Guidelines on Tender Procedures for the Procurement of Pesticides (1994)
13. Guidelines for Retail Distribution of Pesticides with Particular Reference to Storage and Handling at the Point of Supply to Users in Developing Countries (1988)
14. Guidelines for Personal Protection when Working with Pesticides in Tropical Climates (1990)
15. Guidelines on Good Application Practices. Separate documents for ground and aerial applications
16. Guidelines on Procedures for the Registration, Certification and Testing of New Pesticide Application Equipment (2001)
17. Guidelines on the Organization of Schemes for Testing and Certification of Agricultural Pesticide Sprayers in Use (2001)
18. Guidelines on Minimum Requirements for Agricultural Pesticide Application Equipment. Separate documents for different types of equipment
19. Guidelines on Organization and Operation of Training Schemes and Certification Procedures for Operators of Pesticide Application Equipment (2001)
20. Guidelines on Management Options for Empty Containers (2008)
21. Guidelines for the Management of Small Quantities of Unwanted and Obsolete Pesticides (1999)
22. Disposal of Bulk Quantities of Obsolete Pesticides in Developing Countries (1996)
23. Prevention of Accumulation of Obsolete Stocks (1995)
24. Guidelines on Prevention and Management of Pesticide Resistance (2012)
25. Guidelines on Post-registration Surveillance and Other Activities in the Field of Pesticides (1988)
26. FAO/WHO Guidelines on Developing a Reporting System for Health and Environmental Incidents Resulting from Exposure to Pesticides (2010)
27. Guidelines on Monitoring and Observance of the Revised Version of the Code (2006)

health problems unless the spray operators were adequately protected while spraying. The FAO set up a voluntary Prior Informed Consent (PIC) procedure in 1989 to ensure that the importing country had full details of the product and how it should be applied before consent was given for importing it. This led to the Rotterdam Convention in 1998, which established legally binding standards of conduct involving information exchange to enable an importing country to follow the PIC procedure if it needed to use a highly hazardous pesticide. The Convention became effective on 24 February 2004. Under the Convention, there is now guidance on how to monitor and report incidents of pesticide poisoning caused by Severely Hazardous Pesticide Formulations (SHPF). The SHPF Kit can be downloaded from the Rotterdam Convention webpage.

Stockholm Convention

Following Rachel Carson's book *Silent Spring* (Carson, 1962), use of organochlorine insecticides, especially DDT, was prohibited, but considering the wider problem of persistent chemicals in the environment, the Stockholm Convention on Persistent Organic Pollutants (POPs) was adopted on 22 May 2001 and came into force as a global treaty on 17 May 2004. The aim of the treaty is 'to protect human health and the environment from chemicals that remain intact in the environment for long periods, become widely distributed geographically, accumulate in the fatty tissue of humans and wildlife, and have harmful impact on human health or the environment'. This requires a global effort to eliminate or reduce the release of POPs into the environment. In relation to pesticides, a key development has been efforts to ensure that stockpiles and waste contaminated with POPs are managed safely and in an environmentally sound manner, which has involved identifying obsolete stocks, repackaging them where necessary and transporting them to suitable incinerators, following relevant international rules, standards and guidelines. Initially, much attention was given to disposal of dieldrin, which had been strategically stored in areas where locust outbreaks were likely to occur. The ban on organochlorines created the problem of disposal of the existing stocks of dieldrin, where, as previously, any withdrawal of approval required using up existing stocks within two years. One benefit of the ban was the funding of research on a biological mycoinsecticide based on *Metarhizium acridum* that was effective against locusts.

As DDT is covered by the Stockholm Convention, a special case was made for its continued use inside houses for mosquito control. A separate toolkit has been developed to assist with the management of DDT. As of 2016 the following POPs listed in Annexes A and B to the Stockholm Convention have pesticide use: Annex A – aldrin, alpha hexachlorocyclohexane, beta hexachlorocyclohexane, chlordane, chlordecone, dieldrin, endrin, heptachlor, lindane, mirex, pentachlorophenol and its salts and esters, technical endosulfan and its related isomers, and toxaphene. Annex B – DDT.

Montreal Protocol

The Montreal Protocol, signed in 1987 and since amended many times, relates to the use of ozone-depleting chemicals. Banning the use of chlorofluorocarbons (CFCs) used in aerosol spraycans was possible by substituting alternatives. The main problem has been in relation to protection of stored grain and fumigation of soils where methyl bromide was extremely effective. Exemptions have allowed its use in quarantine and pre-shipment of goods, and research has continued to find alternative methods of control.

Minamata Convention

Mercuric chloride had been recommended in the 19th century in a wash with soap applied to the base of apple trees to prevent borers attacking the trees. It was also used as a fungicide to control scab on potatoes. In 1929, it was replaced with mercurous chloride, as it had a lower mammalian toxicity. Mercuric oxide was also commercially available to treat wounds on trees after pruning and control of canker on fruit, rubber and other trees and shrubs. Their use was discontinued, later reinforced by the Minamata Convention to protect human health and the environment from anthropogenic emissions and releases of mercury and mercury compounds. The Convention is named after the Japanese city Minamata, which had suffered a devastating incident of mercury poisoning.

International Standards

International Organization for Standardization

The International Organization for Standardization (ISO) is an independent, non-governmental international organization with a membership of 163 national standards bodies. It began in London in 1946 'to facilitate the international coordination and unification of industrial standards'. Specifications for equipment used to apply pesticides have been developed by working groups and these are considered by national bodies and voted on prior to their publication.

International Union of Pure and Applied Chemistry

The International Union of Pure and Applied Chemistry (IUPAC) was established in 1919 as the successor to the International Congress of Applied Chemistry for the advancement of chemistry. A sub-group is the Committee on Crop Protection Chemistry, which has members from government, academia and industry, and independent consultants, who provide unbiased and authoritative views regarding environmental and human health aspects of crop protection chemistry. IUPAC decides on the common names of pesticides. A Working Party on the Official Control of Pesticides was established by the FAO in 1963 and, subsequently, this was renamed the FAO Panel of Experts on Pesticide Specifications, Registration Requirements and Application Standards, in 1975.

The FAO published the first edition of the *FAO Specifications Manual* in 1971, which has been updated regularly. The scope of the manual was widened in 1998 to bring it into line with the FAO/WHO Joint Meeting on Pesticide Residues (JMPR) to enable JMPR evaluations on pesticide toxicology and residues to be linked to the evaluations of the technical active ingredients. This limited the scope of specifications to technical active ingredients evaluated by the group. The new committee is the FAO/WHO Joint Meeting on Pesticide Specifications (JMPS) composed of scientists acting in their expert capacities and not as representatives of their country or organization. The WHO Pesticide Evaluation Scheme (WHOPES) was set up in 1960 to promote and coordinate the testing and evaluation of pesticides intended for public health use. The WHO also established a committee to develop specifications for equipment for vector control, as there were specific requirements not met by equipment used in agriculture.

As mentioned in the previous chapter, another major problem facing the agrochemical companies is the increase in counterfeit pesticides and sub-standard products, which are sold in both developed and less developed countries. Products are sold that appear to have a label of a 'recognized' pesticide manufacturer but are manufactured by another company and do not conform to the high standards expected. Similarly, there are products marketed with less than the amount of active ingredient stated on the label. Some of these are locally produced and confuse farmers with different trade names and inadequate labelling. Both can result in inadequate control of pests and loss of crop yields. The government regulators have difficulty in prosecuting the dealers of these products as they are not members of a national association, and relevant legislation is not always in place.

Crop Protection Research

The UK depended very much on importation of food, so when DDT became available during World War II, there was considerable interest in its potential, as mentioned in Chapter 1, not only for crops but also for killing mosquitoes. In the 1920s, the Empire Marketing Board, concerned about insect infestation on their dried fruit imported into the UK, asked Prof. Munro, at Imperial College London, to set up a research team. Concerned about doing this research in the college's premises, he requested sufficient funding to purchase a house outside London. A house near Slough was purchased in 1927 and research on fumigation and insecticides, mostly natural pyrethrins, began. In 1938 the Department of Scientific and Industrial Research (DSIR) asked Munro to organize a Grain Infestation Survey and with the outbreak of World War II, DSIR took over Munro's Biological Field Station in 1940, which became the Pest Infestation Laboratory to investigate attacks by insects, mites and fungal pests on harvested crops during all stages of storage, transportation and processing. Dr Page and Lubatti did a tremendous amount of work on fumigation in various situations from barges on the Thames to large warehouses in the docks.

Another scientist was Charles Potter who made critical observations showing that pyrethrum formulations could be residual under the conditions of a darkened warehouse. He wanted a photostable pyrethroid! He designed laboratory bioassay techniques including precision spray applicators to make proper studies of toxicity. After the war, the government retained the Pest Infestation Laboratory. Munro had to find a new field station and selected Silwood Park, near Ascot, which was being vacated by the Ministry of Defence. Training on entomology, crop protection and spray technology (the latter supported by equipment supplied, initially, by the Agricultural Engineers Association) continued at Silwood from 1955 to 2008. At the same time, the WHO used the facilities at Silwood as a collaborating centre for equipment for vector control.

Charles Potter moved to Rothamsted Experimental Station near Harpenden, in charge of the Insecticide and Fungicide Department. His group was active in the introduction of organochlorine and organophosphate insecticides into the commercial agriculture section. Some of the early work on aphid control was done by Michael Way, who subsequently moved to Silwood Park. Potter remained adamant about his aim of getting a photostable pyrethrin insecticide. Michael Elliot, who joined the team, was later able to develop bioresmethrin and, subsequently, permethrin. The government had questioned why a government-supported laboratory was doing what industry should have done, but Potter insisted on the work being continued until the new products were made available to commercial companies.

Although Rothamsted was one of the first research centres established in Victorian times, by 1945 many agriculturally orientated research centres in the UK were involved in aspects of crop protection. Uniquely, the Weed Research Organisation had been set up near Oxford in 1960 as a successor to the Agriculture Research Council Unit of Agronomy. Following the early development of MCPA under the leadership of Prof. G.E. Blackman, a major role of the WRO was independent evaluation of the increasing number of new herbicides being developed by industry. In addition, under Dr E.K. Woodford, the first director, who was succeeded by Dr John Fryer in 1964, the WRO was concerned with understanding the chemistry of herbicides and weed biology.

The WRO was soon regarded internationally as the world centre of excellence for weed science, but the government decided to close it in 1985, despite considerable opposition, to make savings consequent upon a reduction in its income from the Department of Education and Science in 1986/87. To understand the effect of different herbicides, the WRO was particularly concerned with the design of small-plot sprayers (Fryer and Elliott, 1954) and developed a logarithmic sprayer to provide a gradual change in dosage rate across a plot to establish a minimum effective dose (Fryer, 1956). Later, the WRO initiated studies on controlled droplet application of herbicides (Taylor *et al.*, 1976). Long-term studies of the effects of herbicides were ongoing and, undoubtedly, had the WRO not been closed, it would have played a key role in understanding the impact of growing herbicide-tolerant crops. Subsequent work in the UK was divided between Rothamsted and Long Ashton Research Station (LARS), although the latter soon suffered the same fate as the WRO.

The National Fruit and Cider Institute, established in 1903, became the Long Ashton Research Station in 1912, when it became the University of Bristol's Department of Agricultural and Horticultural Research. The unit specialized mainly in research on fruit, and during World War II it helped to produce a home-grown source of vitamin C by developing the blackcurrant drink Ribena. Methods of spraying apple trees were developed by Norman Morgan and his team, who examined ways of reducing the volume of spray that needed to be applied to tree crops (Morgan, 1972) using, in one example, electrically operated spinning discs mounted in front of an axial fan. With the closure of the WRO, LARS became integrated with Rothamsted as the Institute of Arable Crops Research (IACR) in 1986. In 1989 the LIFE (Less Intensive Farming and the Environment) project was initiated by Vic Jordan and stimulated several major integrated crop management research programmes. Eric Hislop and his team now included cereal crops as well as orchards in their research, but further research on tree crops was now concentrated more at East Malling Research Station in Kent, with Long Ashton closing in 2003, 100 years after it had started (Anderson, 2002). The Letcome Laboratory was also closed, with some staff moving to Long Ashton Research Station in 1985, and the site was occupied by Dow Agrosciences Ltd until 2005.

Another closure was the Glasshouse Crops Research Institute (GCRI) at Littlehampton, Sussex, in 1995. It had been established in 1954 to cover research on glasshouse crops, mushrooms, nursery stock and bulbs, by providing better facilities than those available at the Experimental and Research Station at Cheshunt in the Lea Valley (1914–1955). GCRI also took over the work of the Mushroom Growers' Association Research Station at Yaxley, near Peterborough (1946–1954). Much of the work related to crop protection was devising biological methods, including the release of *Encarsia formosa* to control whiteflies and the predatory mite *Phytoseiulus persimilis* to control spider mites. GCRI's expertise became recognized worldwide and later established the efficacy of insect-parasitic nematodes for pest control. Important work was also done on the microbial biopesticide utilizing the toxic protein crystal of *Bacillus thuringiensis* (Burges and Hussey, 1971). Other studies examined the biological efficiency of small droplets (Munthali and Scopes, 1982). GCRI became part of the Institute of Horticultural Research (IHR) with the National Vegetable Research Station at Wellesbourne in Warwickshire, East Malling Research Station in Kent and the Hop Department of Wye College in 1985, and then merged in 1990 with three former Experimental Horticulture Stations – Efford, Kirton and Stockbridge House – to form Horticulture Research International (HRI). All this led to the closure of GCRI in 1995.

Yet another closure was the National Institute of Agricultural Engineering at Wrest Park, Silsoe, in 2006, after 80 years of operation.

The Institute had played a major role in research on tractor mounted sprayers and had a major investment in a wind tunnel to examine the performance of spray nozzles. Studies on electrostatic spraying were also done in the 1970s and 1980s (Marchant and Green, 1982). After closure of the Institute, the spray application unit survived by being linked with The Arable Group (TAG) and the National Institute of Agricultural Botany (NIAB), but became a separate company, Silsoe Spray Applications Unit, in 2016.

Thus from all these closures the research on pesticides used on crops and orchards resides with Rothamsted Research and East Malling Research Station, the two oldest research establishments, which have survived partly because of the way they were originally set up. Rothamsted began in 1843 when John Bennet Lawes, the owner of the Rothamsted estate, appointed Joseph Henry Gilbert, a chemist, as his scientific collaborator and began the classical Rothamsted long-term experiments on Broadbalk field. In 1889, Lawes placed in trust his laboratory and experimental fields at the Rothamsted estate, together with the sum of £100,000, thus creating the Lawes Agricultural Trust (LAT) to ensure the continuation of the agricultural investigations. East Malling began in 1913 with the support of local fruit growers, and, since 2016, became part of NIAB.

Apart from the development of pyrethroid insecticides, studies on pesticides at Rothamsted have been mainly concerned with the problem of resistance, with Roman Sawicki, Paul Needham, Ian Denholm and others contributing to studies both in the UK and overseas. For a short period there was development of an electrostatic sprayer using a spinning disc (Arnold and Pye, 1980). More recently, it has been involved in the 'push–pull' technology and development of GM crops, notably in relation to resisting aphid infestations.

The closure of so many 'independent' research establishments has meant that farmers are increasingly relying on data obtained by commercial companies and some smaller organizations or academics that do research on contract. Even the extent of applied research by commercial companies has been reduced as companies merge and reorganize their programmes. Contributions from industry are discussed later, but in the UK many centres that were active in the 1950s are now closed.

Overseas Research

UK and Commonwealth

The Tropical Pesticides Research Institute (TPRI) was set up near Arusha in Tanganyika (now Tanzania) in the mid-1940s by the UK government, when DDT and other new insecticides were available to do research on major tropical pest problems. This was in conjunction with a similar unit based at Porton Down in the UK. TPRI worked under colonial government, East Africa Common Services Organization (EACSO), East Africa

Community (EAC) and, currently, the Tanzanian government through Parliament Act No. 18 of 1979.

Early work included studies on controlling tsetse flies with aerial sprays using war-surplus aircraft, applying sprays and coarse aerosols of DDT and HCH. These sprays were ineffective against tsetse flies because the forest canopy was too thick to allow adequate penetration of large spray droplets, resulting in sub-lethal dosages on the vegetation below the canopy (Hocking *et al.*, 1953; Hocking and Yeo, 1956). Applications of coarse aerosols, although giving higher mortalities of tsetse than did sprays, were not completely successful because most of the insecticide was blown away from the target area. With the development in recent years of improved aerosol-dispensing equipment, more toxic insecticides and a better understanding of the fate of aerosol particles in relation to prevailing weather conditions, appreciable advances in tsetse control were made (Lee *et al.*, 1969). The Institute continues to conduct research into tropical pests affecting plants, livestock and human health.

In the 1960s and 1970s, assistance was given by staff at Porton to other Colonial Office projects in Africa, including the Cotton Pest Research Scheme in the Federation of Rhodesia and Nyasaland, and later in Malawi, set up by Eric Pearson, Director of the Commonwealth Institute. Dating back to 1888, the Imperial Institute was established by royal charter from Queen Victoria, and its name was changed to the Commonwealth Institute in 1958. The Imperial Institute of Entomology, under Sir Guy Marshall (1871–1959), began publishing the *Bulletin of Entomological Research* in 1910, and the *Review of Applied Entomology* in 1913, a major publication in the days before the internet. The name became the Commonwealth Institute of Entomology in 1947 and later became part of CABI (Commonwealth Agricultural Bureau International), which had its roots in the 1930s.

Universities in the USA

In the USA, universities in each state provided research and extension services to farmers and other users of pesticides. Some of these specialized in certain areas, thus the University of California, at Davis, remains involved in developments in aerial application. Prof. Norman Akesson and Wes Yates provided publications for agricultural and vector control with aircraft and, recently, Ken Giles has worked with unmanned aerial vehicles (UAV). At Berkeley, Prof. Ray Smith and others played a leading role in the development of integrated pest management, attending meetings in Rome to get the concept adopted globally. At Riverside, attention was more on the problem of resistance to pesticides. Elsewhere in the USA, research at Texas A&M, where Fred Bouse and Jim Carlton were active, spanned work on crops such as cotton, development of aerial sprays and use of electrostatic sprays, as well as efforts to release natural enemies and irradiated screwworm flies from aircraft. At Nebraska, new laboratories were constructed in 2012 to continue work centred on the application of herbicides over four decades, while in Oregon a group examined manually carried equipment as used further south in central and South America, with Alan Deutsch coordinating an IPM newsletter for many years. In Ohio, there were laboratories at Wooster for the US Department of Agriculture (USDA) and also at the university, where a research group was initiated by Frank Hall, the Laboratory for Pest Control Application Technology (LPCAT), which was mostly concerned with basic information on droplet size and their distribution, but which also did work on orchard crops.

Other countries

Most countries within the EU have one or more laboratories that are responsible for different aspects of pesticide use. These enable chemical analyses of products or to determine residues in foods, while engineering organizations do work on pesticide application. In Australia, the Agricultural College at Gatton, now part of the University of Queensland, organized pesticide application training courses and conducted research programmes.

Consortium of International Agricultural Research Centers (CGIAR)

CGIAR was set up as a consortium for international agricultural research. It is unusual in that it is not part of an international political institution, such as the UN or the World Bank, but is supported with funds from its members. Members of CGIAR include governments, namely those of the USA, Canada, UK, Germany, Switzerland and Japan, together with various institutions, philanthropic foundations, including the Ford Foundation, and organizations such as FAO, IFAD and development banks. Its aim has been to advance agricultural research for development and to ensure a food-secure future. The research centres concerned with major crops (see Box 10.1) have been mainly responsible for developments in breeding new crop varieties, with biological control the preferred option. Pest management and pesticide use has not featured prominently in their programmes but some centres have used pesticides. IRRI, in the 1970s, applied pesticides at exceedingly high volumes using a lance on a hose with the pump operated on a tractor at the end of an irrigated plot. At that time, brown plant hopper, *Nilaparvata lugens*, was a problem, partly because there were overlapping crops in the area. As the insect was feeding at the base of stems, it is doubtful that the very diluted spray had much effect, but luckily a more resistant variety was selected. Spraying was also not efficiently carried out at ICRISAT, where a large tractor was used on some crops, spraying both crop and inter-row.

Box 10.1. Research centres that operate as part of the CGIAR system.

AfricaRice, Cote D'Ivoire
Bioversity International, Rome
Center for International Forestry Research (CIFOR), Indonesia
International Center for Agricultural Research in the Dry Areas (ICARDA), Lebanon
International Center for Tropical Agriculture (CIAT), Colombia
International Crops Research Institute for the Semi-Arid Tropics (ICRISAT), India
International Food Policy Research Institute (IFPRI), USA
International Institute of Tropical Agriculture (IITA), Nigeria
International Livestock Research Institute (ILRI), Kenya
International Maize and Wheat Improvement Center (CIMMYT), Mexico
International Potato Center (CIP), Peru
International Rice Research Institute (IRRI), Philippines
International Water Management Institute (IWMI), Sri Lanka
World Agroforestry Centre (ICRAF), Kenya
WorldFish, Malaysia

Extension Services

In the early days of crop protection, the advice provided to farmers came from universities, and in the UK agricultural colleges were set up in different areas. The Royal Agricultural College at Circencester was set up in 1845 and was granted its royal charter shortly afterwards. It was followed by Writtle College in 1893 and Harper Adams University College in 1901, the latter becoming a university in 1996. It was noted for a diploma course in crop protection, enabling students to be employed in various sectors of the agrochemical industry in the UK. Many counties had their own agricultural college, many of which have since diversified to embrace environment and rural development. Further north, the West of Scotland Agricultural College was formed in 1899, the East of Scotland Agricultural College in 1901 and the North of Scotland Agricultural College in 1904, and these amalgamated to form the Scottish Agricultural College in 1990. The county colleges provided courses for the agrochemical companies under the British Agrochemical Industry Scheme, later referred to as BASIS, which enabled distributors of pesticides to keep their certification up to date. Similar training is not generally available in many countries, where a farmer needing an insecticide may be sold a fungicide!

With the need to increase crop production post-World War II, a National Agricultural Advisory Service (NAAS) was established in 1946 as part of the Ministry of Agriculture, Fisheries and Food (MAFF). Specialist advisers in plant pathology, entomology, soil and other subjects were able to advise farmers and growers how to maximize their output. In 1971, NAAS became the Agricultural Development Advisory Service (ADAS) and became an executive agency of MAFF in 1992, prior to being privatized in 1997.

Extension services in other countries are either linked to agricultural research institutes or, as in the USA, to universities. Sadly, in many countries the role of extension services has not been supported by governments, so farmers have had to rely on commercial consultants and those selling pesticides rather than have independent advice based on local research.

Agrochemical Companies

Earlier in this book, it was mentioned that William Cooper started selling chemicals in 1843 and his business prospered after he died in 1888. The business was later run by Sir Richard Ashmole Cooper, but in 1925, due to poor health, he decided that the company should be amalgamated with McDougall and Robertson Ltd. He also set up the Cooper Research Laboratory in Berkhamsted. Cooper McDougall and Robertson later produced an anti-louse powder (AL63) during the war, which contained DDT, as many men had died of trench fever and typhus during World War I. The company was taken over by Wellcome in 1959, and in 1992 the Wellcome Foundation sold its environmental health business, including the Berkhamsted site, to the French company Roussel Uclaf. By 1995, Roussel's major shareholder, Hoechst, joined forces with Schering to form AgrEvo, but they closed the Berkhamsted site in 1997. Earlier, in 1937, Cooper McDougall and Robertson formed an association with ICI to form the Plant Protection Division.

In 1938/39, Sir Guy Marshall (Fig. 10.1), the director of the Imperial Institute of Entomology, teamed up with Dr Walter Ripper (Fig. 10.2) to form Pest Control Ltd. During the 1930s, Ripper had developed a simplified method of vaporizing nicotine applied to a field crop under a

Fig. 10.1. Sir Guy Marshall. (From Fisons, 1976.)

Fig. 10.2. Dr Walter Ripper. (From Fisons, 1976.)

Fig. 10.3. Walter Ripper walking behind a tractor applying nicotine under a layer of sacking. (From Fisons, 1976.)

'portable' tent dragged slowly behind a tractor to control aphids in crops (Fig. 10.3), and later a no-drift spray boom (Fig. 10.4). Pest Control Ltd began operations in a disused garage near Hauxton Mill, and by 1943 they were making pesticides and tractor sprayers as well as doing contract spraying. Initially, they applied 'Burgundy' mixture, which was easier to mix than Bordeaux, to treat potato crops. During World War II they also sprayed paint on airfields to disguise areas of grass as fields of crops between hedgerows. Pest Control Ltd developed the use of sulfuric acid to destroy potato haulms without needing to dilute it. Modification of sprayers enabled improved spraying of orchards, and the company also started to manufacture the weedkiller DNOC to replace supplies that had been obtained from France.

Fig. 10.4. Boom with wind shield developed by Pest Control Ltd. (From Fisons, 1976.)

Pest Control Ltd started to provide scientists to do research under contract, but this was not sustained as there was an inevitable conflict of interest when the research findings did not support the use of a treatment most profitable for the companies' chemical operations. This was especially true when the company was taken over by Fisons, who wanted to produce chemicals for the widest possible market instead of solving specific pest control problems.

In 1952, Pest Control Ltd bought the Chesterford Park estate to develop its research activities, but in 1954 Fisons Ltd took over the company. Fisons was a company that dated back to the 17th century, with a flour mill in Birmingham and later a malting business, eventually producing sulfuric acid and fertilizers. The formation of Fisons Pest Control added agrochemicals. Key scientists at Chesterford Park were Dr Edson, a medical toxicologist, Dr Greenslade and Ron Amsden, who had experience of aerial spraying in the Gezira, Sudan, where DDT was sprayed on cotton to control jassids using S51 and, later, Hiller helicopters. In 1962, he wrote that 'every effort should be made to produce a spray with very few drops greater than 120 microns and very few smaller than 80 microns'. The aim was to produce a wider swath and more uniform coverage, but recognizing that the droplet spectrum arriving at the crop is not the same as that leaving the nozzle, they did early research on anti-evaporation formulations (Amsden, 1990). Amsden also pointed out that certain leaf surfaces, such as peas, bananas and cabbages, tend to reflect droplets larger than 100 microns, a factor generally ignored in so many spray programmes, applying pesticides diluted in large volumes of water and using large droplets to avoid spray drift. Later the company helped British Rail develop a spray train to kill weeds on the tracks (Fig. 10.5). The company also had interests in the research and control of cotton pests in the Gash delta and Nuba Mountain region of Sudan, as well as pest control in the Elgin apple-growing region of South Africa.

Jesse Boot, who founded Boots and Co. Ltd, the chemists, in 1883, began to establish an agricultural division much later, in 1929, but it was only in 1947 that new laboratories for horticultural research were completed at Lenton. One interesting product was a plant growth regulator that could increase yields and reduce lodging of plants. In 1980, this part of Boots joined Fisons to form FBC Ltd, but the company was soon acquired by Schering in 1983 and then became part of AgrEvo in 1995.

In the post-DDT era, there was a large increase in companies producing pesticides. Some had been established earlier in the production of dyes, paints and other chemical products, but quickly added pesticides. Geigy, established in 1758, produced pharmaceutical drugs and dyes and patented DDT in 1940, developing simazine herbicide later. In 1969, Ciba amalgamated with Geigy to form Ciba Geigy. Much later, they acquired Sandoz, which had acquired Zoecon in 1984 and Velsicol a year later, and

Fig. 10.5. The first spray train developed by Pest Control Ltd. (From Fisons, 1976.)

formed Novartis in 1996. ICI Plant Protection Division had developed the technique of direct drilling, based on the use of paraquat (Gramoxone) as a contact herbicide since 1961, among other major developments, and also stimulated interest in electrostatic spraying with the development of the Electrodyn sprayer (Coffee, 1979), which it decided not to pursue. In 1987 it had acquired Stauffer and became Zeneca in 1993. While the pharmaceutical part of Zeneca merged with the Swedish Astra to form AstraZeneca, the crop protection part of Zeneca linked with the agrochemical part of Novartis, which had acquired ISK Biosciences in 1998 to form Syngenta in 2000. Syngenta was absorbed by a takeover by ChemChina to create an agrichemical powerhouse.

Among the oil companies, Shell Chemical Company played a key role in developing the cyclodiene insecticides aldrin, dieldrin and endrin in the 1940s. Shell were leaders with soil fumigants 1,2-dibromo-3-chloropropane (Nemagon) and 1,3-dichloroprpene (D-D), developed by Dow, later produced mevinphos (Phosdrin), in 1953, and marketed diclorvos (Vapona), originally developed by Ciba in 1959. Along with other companies that decided not to stay in developing pesticides, Shell, which had acquired CelaMerck, sold their UK research base to American Cyanamid in 1993. This was then acquired by American Home Products a year later with the crop protection part being merged with BASF in 2000. The USA part of Shell was bought by DuPont in 1985.

Bayer Crop Science, formed in 2002, was the result of a series of mergers. Bayer combined with Aventis, a company that had developed from a merger between AgrEVo and Rhône-Poulenc, which had already acquired Union Carbide in 1987, having acquired Amchem in 1977. Rhône-Poulenc had previously set up a separate UK company called May & Baker in 1927, which was also absorbed into Aventis. As mentioned earlier, AgrEvo was the product of a merger that included Hercules, Fisons, Boots agrochemical division and Upjohn, prior to 1985.

Elanco sold out its interests in agrochemicals to Dow after a short period of a merger and Dow AgroSciences was formed in 1997. It later acquired Rohm and Haas and the seed/molecular biology company Mycogen. In the USA, FMC, founded in 1883, was originally known as the Bean Spray Pump Company, as John Bean had developed the first piston pump insecticide sprayer. Later, as the company grew, it changed its name to the Food Machinery Corporation, and later the Food Machinery and Chemical Corporation, which eventually became known as FMC Corporation. It has recently bid to take over part of DuPont's crop protection business.

Sumitomo has become the largest Japanese agrochemical company and it acquired Valent BioSciences, while it also linked with a domestic company in 2002 to create the Sumitomo Chemical Takeda Agro Company, in order to have a stronger presence in the local Japanese market.

Once Monsanto had started marketing the genetically modified crops and acquired seed companies, the other major companies also started to buy seed companies, and, more recently, to add in bioscience companies as the EU put integrated pest management into the pesticides legislation. In 2012, Syngenta acquired Pasteuria Bioscience, while in the same year BASF purchased Becker Underwood, which had developed a strong market for entomopathogenic nematodes and mycoinsecticides as part of their Functional Crop Care in the Crop Protection division, and Bayer purchased AgraQuest, a 16-year-old biopesticide company, and the Israeli company Agrogreen. However, in 2017, Bayer was able to purchase Monsanto, which, with few pesticides, had become the major company developing genetically modified crops. Also in 2017, Dow Chemical and Du Pont agreed to merge and have three separate units to concentrate on agriculture, material science and the production and sale of speciality products.

Alongside the major R&D agrochemical companies, there has been a vast increase in generic pesticide companies who market the off-patent products. Major generic companies include:

(1) Adama, formerly Makhteshim, which is now part of ChemChina;
(2) Nufarm, which has manufacturing and marketing operations in Australia, New Zealand, Asia, Europe and the Americas;
(3) United Phophorous Limited, incorporated in India in 1969 but now better known as UPL, and is the largest pesticide manufacturer in India, marketing products globally;
(4) Cheminova, which has been part of FMC since 2015; and
(5) Sipcam, an Italian multinational company, private and independent, specializing in the production, marketing and sales of plant protection products and chemical intermediates. It has marketed globally with subsidiaries in some 12 countries.

Pest Control Companies

Most countries have privately owned pest control companies that generally serve towns and cities by providing trained personnel to control pest problems such as rats, cockroaches, wasps and bed bugs, which occur in homes, hotels, offices and other areas. In the UK, the British Pest Control Association was set up in 1942 as the Industrial Pest Control Association to preserve the nation's food stocks in time of war and to maintain public health through stewardship of limited stocks of pyrethrum. The first British Pest Conference was held in 1963, and in 1969 the association became the British Pest Control Association. This coincided with the demise of the voluntary Pesticides Safety Precautions Scheme, so the Association developed its own training course, codes of practice and conferences with government-recognized examinations and qualifications. In 1974, the Association linked with similar groups in Belgium, France, Holland and Spain to form the Confederation of European Pest Control Associations (CEPA) to defend their interests at a European level. In 1980, the BPCA Training and Certification Scheme for pest control and fumigation operators was set up, with the first five-day course held at Aston

University in Birmingham. After becoming a private limited company in 1982, it awarded its first Proficiency Certificate in 1984. The conferences were replaced by an exhibition – PestEx – in 1995, and have continued biannually. BPCA has now organized a probationary scheme to assist new pest control companies.

Crop Protection Organizations

CropLife International

In 1967, the International Group of National Associations of Manufacturers of Agrochemical Products (*Groupement International des Associations Nationales de Fabricants de Produits Agrochimiques* (GIFAP)) was founded. In 1996, GIFAP was renamed the Global Crop Protection Federation (GCPF), and in 2001 it was renamed CropLife International to represent the plant science industry. It is based in Brussels. The major R&D companies BASF, Bayer, Dow AgroSciences, DuPont, FMC, Monsanto, Sumitomo and Syngenta support the organization in providing information on behalf of the whole industry. In addition to CropLife, the industry has regional associations, such as the European Crop Protection Association. Recently, the commercial companies marketing biopesticides set up the International Biopesticide Manufacturers Association (IBMA). There is also a Biological Products Industry Alliance (BPIA) that includes other biological products such as biostimulants.

British Crop Production Council (BCPC)

In some countries, a national association has been established to organize conferences to disseminate the results of the rapidly growing field of agrochemical research and development and to provide independent information to those involved in pesticide use. In the UK there were originally two organizations interested in the new development of pesticides – the British Weed Control Council and the British Insecticide and Fungicide Council, which merged in 1967 to form the British Crop Protection (now Production) Council (BCPC) under the presidency of Sir Frederick Bawden who was Director of Rothamsted Experimental Station. Early members of BCPC included the forerunners of Rothamsted Research, the Association of Agricultural Contractors (NAAC) and the Crop Protection Association (CPA).

Annual conferences, combined with an exhibition, were held at Brighton, originally alternating the weed and pest/disease topics, and these attracted an international audience of scientists from industry and academia as well as those involved with regulation. In the early conferences, the conference dinner was followed by an invitation to the local nurses to come for a dance. Industry made full use of the large attendances to hold separate meetings alongside to discuss and promote new products. BCPC published the proceedings of the conference papers and also produced the *UK Pesticide Guide* for growers and advisers; and in 1968 it produced the *Pesticide Manual*, a globally recognized reference book, now being superseded by a fully searchable online version. This was followed by the *BioPesticide Manual* and a *Manual of GM Crops*.

Meetings of small BCPC Expert Working Groups reviewed developments in the industry and this led to the publication of technical training handbooks – *Field Scale Spraying, Small Scale Spraying, Using Pesticides, Spreading Fertilisers and Applying Slug Pellets* and *The Safety Equipment Handbook*. In 1985, the BCPC spray quality system was introduced and later developed as an International standard. In the digital age, online resources were set up, including the weekly *BCPC News*, disseminating links to global news items from a wide range of publications. Since 1985, BCPC has awarded medals to those who have made a significant contribution to UK crop protection and production.

BCPC has continued to provide a platform for discussions aimed at improved and better targeted agrochemicals, integrated pest management and biotechnology, and in 2000 became the British Crop Production Council, reflecting its broadened approach to promoting the science and practice of sustainable crop production. The working groups deliver annual reviews of the latest applied research findings to invited audiences.

The conference moved to Glasgow in 2003, but outside events led to a return to Brighton on a smaller scale, with emphasis on EU regulations. This focus will continue while the UK is exiting the EU as it provides an opportunity to reemphasize science as the basis of a UK crop production regulatory regime.

In the modern world, where the rhetoric of pressure groups can influence policy more than evidence-based judgement, the mission of BCPC and its independent status is increasingly needed to defend science in crop production and in its regulation.

Association of Applied Biologists (AAB)

As in many countries there are scientific societies for entomologists and other disciplines. In the UK, the Association of Applied Biologists has a pesticide application group that, since 2000, has organized an international conference on application technology every two years. Other groups have organized annual meetings on various subjects within the context of crop protection.

International Association of Plant Protection Societies (IAPPS)

The first International Plant Protection Congress was held in Louvain, Belgium, in 1946. Initially, the control of insects was the major theme, but at the 9th Congress in Washington, DC, in 1979, the programme was organized by a multidisciplinary group consisting of plant pathologists, entomologists, weed scientists, nematologists and chemists. Programmes have continued to reflect the integration of these disciplines. BCPC hosted the Congress in 1983, with Professor Leonard Broadbent as president. Subsequently, at the conference in Jerusalem in 1999, the International Association for the Plant Protection Sciences (IAPPS) was formed. IAPPS now coordinates the major conference held every four years. Its newsletter is published in the journal *Crop Protection*. Distinguished scientists in crop protection have been given an award at these conferences.

Apart from IAPPS, individual societies covering the various scientific disciplines within crop protection hold specialist and international meetings to discuss developments in research.

European Plant Protection Organization (EPPO)

The European Plant Protection Organization, set up in 1951, has its headquarters in Paris. It is the regional plant protection organization for Europe and now is supported by 51 member governments, which include countries around the Mediterranean and some in central Asia. It has working groups on phytosanitary regulations and plant protection products. Prior to the more recent EU directives, EPPO sought methods to harmonize the evaluation of pesticides for specific pests and crops and published advisory bulletins. EPPO also organized conferences to bring together crop protection scientists in Europe, and published a bulletin with information from conferences and other relevant topics.

International Organisation for Biological Control (IOBC)

The IOBC was established in 1955 to promote environmentally safe methods of pest and disease control in plant protection. It has six regional sections. Members of the West Palaearctic Regional Section (WPRS) are individual scientists and governmental, scientific or commercial organizations from 24 countries of Europe, the Mediterranean region and the Middle East. It was established to foster research and practical application, organizes meetings and symposia, offers training and information, especially on biological methods of control, and includes chemicals within an integrated pest management context.

Malaysian Plant Protection Society (MAPPS)

The MAPPS was established in 1976 and aims to serve as a platform to discuss and generate knowledge pertaining to plant protection following the same aims as the BCPC. Conferences and training courses are held to improve the use of pesticides and to protect the environment.

References

Amsden R.C. (1990) Reducing the evaporation of sprays. *Agricultural Aviation* 4, 88–93.

Anderson, H. (2002) Long Ashton Research Station: one hundred years of research. *Pesticide Outlook* 13, 214–217.

Arnold, A.J. and Pye, B.J. (1980) Spray application with charged rotary atomizers. *British Crop Protection Council Monograph* 24, 109–125.

Burges, H.D and Hussey, N.W. (eds) (1971) *Microbial Control of Insects and Mites.* Academic Press, London.

Carson, R. (1962) *Silent Spring.* Houghton Mifflin Co., Cambridge, Massachusetts, USA.

Coffee, R.A. (1979) Electodynamic energy – a new approach to pesticide application. *Proceedings of the British Crop Protection Council Conference – Pests and diseases.* 777–789. British Crop Protection Council, Farnham, UK.

Fisons (1976) *The History of Pest Control.* Fisons Ltd.

Fryer, J.D. (1956) A small-scale logarithmic sprayer. *Proceedings of the 3rd British Weed Control Conference*, 585–590.

Fryer, J.D. and Elliott, J.G. (1954) Spraying equipment for the experimental application of herbicides. *Proceedings of the 2nd British Weed Control Conference*, 375–388.

Hocking, K.S. and Yeo, D. (1956) Aircraft applications of insecticides in East Africa. XI. Applications of a coarse aerosol to control *Glossina morsitans* Westw. at Urambo, Tanganyika, and *G. morsitans* Westw.,and *G. pallidipes* Aust. in Lango County, Uganda. *Bulletin of Entomological Research* 47 631–644.

Hocking, K.S., Parr, H.C.M., Yeo, D. and Anstey, D. (1953) Aircraft applications of insecticides in East Africa. *Bulletin of Entomological Research* 44, 627–631.

Lee, CW., Coutts, H.H. and Parker, J.D. (1969) Modifications to micronair equipment and assessment for fine aerosol emission in tsetse fly control. *Agricultural Aviation* 11, 12–17.

Marchant, J.A. and Green, R. (1982) An electrostatic charging system for hydraulic spray nozzles. *Journal of Agricultural Engineering Research* 27, 309–319.

Matyjaszczyk, E. and Sobczak, J. (2017) Common EU registration rules and their effects on the availability of diverse plant protection products: a case study from oilseed rape and potato in 5 Member States. *Crop Protection* 100, 73–76.

Morgan, N.G. (1972) Spray application in plantation crops. *Pest Articles and News Summaries* 18, 316–326.

Munthali, D.C. and Scopes, N.E.A. (1982) A technique for studying the biological efficiency of small droplets of pesticide solutions and a consideration of its implications. *Pesticide Science* 13, 60–63.

Taylor, W.A., Merritt, C.R. and Drinkwater, J.A. (1976) An experimental tractor-mounted machine for applying herbicides to field plots at very low volumes and varying drop sizes. *Weed Research* 16, 203–208.

11 Pesticides – the Future

The application of pesticides has played a key role in crop protection during the Green Revolution, allowing farmers to reap higher yields from new varieties. Yield increases have tended to plateau in recent years, yet the human population continues to increase and is expected to reach 9 billion by 2050, thus requiring an expanding supply of food. Certain antipesticide organizations argue that we should favour growing 'organic' crops, but with climate change and the continuing spread of plant diseases and insect pests, largely due to global trade, the need for crop protection will continue if people are to be adequately fed. In the foreseeable future, pesticides will have to play a role despite mounting problems of increasing resistance of pests to existing products, greater concerns about the environment and opposition in many countries to new technology, including the growing of genetically modified crops.

Resistance Management

A change is needed to extend the practical effective lifetime of pesticides. At present, once a specific chemical is shown to be effective and its use increases, the cost of the formulated product tends to decrease, allowing an expansion of its use. The consequence of this has been overuse, and inevitably, pests become resistant to it, as discussed earlier. The response so far has been mainly a search for a new chemical with a different mode of action, with few examples of a deliberate rotation of products within an area to limit selection pressure for resistance. The cost of searching for new pesticides with a novel mode of action in a world requiring ever more testing of effectiveness without harming the environment has increased to such an extent that fewer companies will be able to invest in new products. This is already apparent in the merging of companies and by the adoption of a policy of integrated pest management aimed at the need to find and develop biopesticides. While biopesticides present a different agenda for evaluation, they, too, will require significant investment to achieve effective commercial products, which can be easily stored and applied by farmers.

Following the discovery of DDT, new pesticides were developed quite rapidly, but regulatory requirements and increased costs to develop a new molecule has transformed the situation. In an endeavour to improve the situation for control of vectors of human diseases, the Gates Foundation has financed the Innovative Vector Control Consortium, which has assisted with evaluating potential insecticides, primarily for mosquito control. This was largely because industry had not invested in looking for insecticides for vector control as they represented only 1% of the pesticide market. So far, some insecticides used in agriculture have been reformulated, thus pirimiphos methyl is now used as a micro-encapsulated formulation to increase its persistence on wall surfaces when applied as an indoor residual spray. Other new products are under development, but details have yet to be released.

New molecules have come from knowledge of botanical insecticides, the pyrethroid insecticides being an example of development from pyrethrins, but this route depends on whether other suitable botanicals can provide suitable starting points. The use of neem, from the tree *Azadiracta indica*, as a botanical insecticide achieved variable results depending how the neem was extracted from leaves or other parts of the tree, but the active agent in neem, azadirachtin, is a complex molecule and has not been developed, presumably due to its production costs and its rapid breakdown. Not all botanical extracts are safe; for example, nicotine from tobacco is a very toxic insecticide, although industry has been able to develop several neonicotinoids, which have attracted growing concerns over their toxicity to bees. Thus, particular care is needed to ascertain the mammalian toxicity of extracts of other botanicals that have been shown to be toxic to pests in many laboratory studies, prior to being marketed for field use. There is also the problem that there is variation in the amount of insecticidal chemical that is present in a plant, as shown in fish poison bean, *Tephrosia vogelii* (Belmain *et al.*, 2012). In the foreseeable future there will certainly be more research on biopesticides, which will play an increasingly important role in IPM.

Novel Biocontrol Agents

Spider venom

Venom from spiders has provided a source of stable insecticidal proteins that can cause insect paralysis through modulation of ion channels, receptors and enzymes. Research on these has characterized insecticidal toxins that target novel sites of action in insects (King and Hardy, 2012; Windley *et al.*, 2012). Nakasu *et al.* (2014) reported using fusion protein technology to link insecticidal peptides to a plant lectin 'carrier' protein to orally deliver the toxin as a biopesticide. Thus the creation of a novel biopesticide by fusing Australian funnel-web spider (*Hadronyche versuta*) venom (versutoxin, Hvla) with snowdrop flower (*Galanthus nivalis*) proteins has been shown to be active against agricultural pests while leaving honey bees unharmed. A range of novel products aimed at specific pests has now been approved in the USA, Ullah *et al.* (2015) reported encoding Hvla toxin in cotton and tobacco plants, but when tested with *Heliothis virescens*, the toxin expression was much lower in cotton compared with 100% larval mortality on tobacco.

Wang and St Leger (2007) showed that a scorpion neurotoxin could increase the potency of a fungal insecticide. Subsequent studies have shown that genetically modified fungi can now be obtained to increase their effectiveness as biocontrol agents (Bilgo *et al.*, 2017). This has been demonstrated using transgenic fungi (*Metarhizium*) applied to sheets on which mosquitoes can rest, which resulted in effective control of malaria within five days, indicating that malaria transmission could be effectively reduced. However, a cautious public may prefer a biopesticide without any genetic modification.

Neuropeptides

In Europe, as a result of a policy of integrated pest management, using pesticides as a last resort, a project was set up to look for novel biocontrol agents for insect pests from studies on neuroendocrinology. Neuropeptides are small, protein-like molecules (peptides) used by neurons to communicate with each other. These neuronal signalling molecules influence the activity of the brain and the body in specific ways. The nEUROSTRESSPEP project, running for four years from 2015, is looking at insect neuropeptides, which are quite different from human neuropeptides. Not all insects use the same signals, so the plan is to identify neuropeptides that are shared by agricultural pests but not by beneficial insects, aiming to design new chemicals that resemble the structures of these peptides (Davies, 2017).

Spray Drift

The movement of pesticides away from a treated target area downwind as spray droplets or vapour has become a very important issue as there has been increasing concern about the potential effects of spray drift reaching non-target organisms, which may be very sensitive to small amounts of pesticide. The most spectacular effects have been mostly due to damage to sensitive plants caused by herbicides, such as 2,4-D and dicamba, especially when vapour moves from spray deposits within a crop area in much the same way as the downwind movement of spray droplets. As indicated in Chapter 9, the extent of exposure of people to pesticides as spray drift is somewhat limited, although there have been occasions when habitations within plantations have been oversprayed by aircraft. Nevertheless, there is a problem due to the need to minimize drift by using larger droplets, yet achieve good coverage of target areas with pests, which necessitates smaller droplets. Improved formulations may provide some protection from spray drift combined with careful nozzle selection and operational factors.

This problem is likely to increase as farmers are tending to travel across fields with their sprayers at faster speeds and with wider booms, which may be set higher above the crop. Adoption of ultra-low-volume spraying has not been welcomed by registration authorities because of perceived greater risk of absorption of an oil-based formulation through the exposed skin of operators. Using ULV sprays by spraying downwind of the operator in Africa on cotton was successful in the field, but the agrochemical industry has taken little interest in application technology other than checking that their products can be sprayed through standard equipment with hydraulic nozzles. Forays into special applications, such as use of electrostatic spraying with the Electrodyn, were curtailed to some extent, due to problems in developing suitable formulations for a range of different pesticides. Perhaps the improved efficacy obtained with electrostatically charged droplets that reached the target foliage was another factor as lower dosages could be applied. Whether a ULV formulation could be developed with a very fine particle suspension in a vegetable oil has not been fully explored, but could be important as a means of reducing loss of deposits following rain. Deposits achieved with pesticides formulated to mix in water are undoubtedly exposed to the effect of rain, with run-off ultimately draining from fields and contaminating water in rivers.

Drones

Pesticide application technology has not changed significantly, when we realize that pesticides were mixed in water and sprayed through hydraulic nozzles in the 19th century. However, with the development of unmanned aerial vehicles – drones capable of flying accurately over fields, controlled by advanced global positioning systems (Lan *et al.*, 2017) – it is speculated that instead of taking thousands of litres across fields and compacting the soil, sprays should be applied with a drone. Drones have already been employed to treat rice fields in Japan for two decades. Reports indicated that by 2010, 30% of Japanese rice fields were treated using a RMAX drone-copter or a robotic rival, while the area treated with manned helicopters decreased. Further development of technology has enabled infrared and thermal cameras to be fitted to drones, which can scan crops at night to detect insects congregating in harmful numbers and target a spray, thus reducing the area sprayed with pesticide. The agri-drone has become a precision tool in south-east Asia (He *et al.*, 2017), while using similar systems in Europe and North America is only just surfacing as a practical way ahead. Many of the drones now in use are small and carry a camera, but ideally a drone is needed with a capability to lift a larger volume of spray to minimize the number of times it requires refilling to spray large farms (Figs 11.1, 11.2).

Fig. 11.1. Unmanned aerial drone with camera for collecting data on crop. (Photo courtesy of Sensat, used with permission.)

Fig. 11.2. Drone spraying rice. (Photo courtesy of Yamaha, used with permission.)

Targeting specific insect swarms may be possible for certain pests, but if the drone is to be effective for weed and disease control, development of an ultra-low-volume application system spraying at relatively low speeds (20–30 km/h) low over a crop, equivalent to a tractor spray boom, is needed. In conjunction with aerial surveys, it should still enable distribution of the pesticide to be targeted where weed, disease or insect pests are located. More accurate atomization to achieve a narrow droplet spectrum, probably using rotary atomization, to optimize downward trajectory of the spray with evaporation retardant should minimize spray drift.

Interestingly, instead of spraying herbicides, there is also the idea that it might be possible to use a drone equipped with a laser that could use imagery from drone-mounted cameras able to respond to a very wide range of wavelengths, analyse the data to provide weed identity data and damage the weeds without affecting adjacent plants, but this will depend on the nature of the target and its surroundings.

Seed Treatment

Getting a crop established once there is sufficient rainfall is beset with many problems, especially for small-scale farmers who have needed family help to weed their food crops during the first weeks of plant growth. The development of herbicide-tolerant crops, discussed later in this chapter, allows a farmer to delay using a herbicide if early rains are erratic and adversely affect initial crop growth, but seed treatment with a systemic insecticide may be crucial to prevent the crop suffering from infestations of early season sucking pests such as aphids. By using a seed treatment, the need to spray young plants, when much of the spray will be deposited on the intercrop area, is avoided. With a system of slow release formulations on seed, the period of protection can be lengthened and minimize leaching of chemicals in the soil.

Genetically Modified Crops

Ever since man started to grow crops there was a selection of seeds to sow the following season's crops. In the UK, Thomas Knight (1759–1838) was one of the earliest plant breeders to deliberately select plants with better qualities. In the late 19th century Garton's Agricultural Plant Breeders was formed and commercialized new varieties developed by cross-pollination. Using hybrids and continuing selection, plant breeders endeavoured to select for higher-yielding varieties and with resistance to diseases and insect pests. It is well known that many, if not all, plants have some biochemical system to protect them from invasion of insect pests and diseases and man has exploited this characteristic of wild plants by the extraction of pyrethrins and subsequent development of pyrethroid insecticides. Similarly, this has occurred with nicotine from tobacco leading to the neonicotinoid insecticides.

The most important change has been to utilize molecular biology to select or, in the case of genetic modification, to insert desirable traits into plants, leading to the development of genetically modified crops since 1983. Initially, studies used tobacco plants, but the interest was on food crops. The first genetically modified food crop was tomatoes to increase the shelf life of the fruit by inserting a gene that delayed ripening. In 1994, the Flavr Savr tomato was introduced and was also used in production of tomato paste, but the main emphasis was on developing crops that were easier to grow. Recently, a new technology called CRISPR has been developed to take advantage of bacterial systems to simplify genetic editing, allowing for easier development of genetically modified (GM) crops. There is the potential to improve a plant's resistance to key diseases, both due to fungi and viruses as well as to insect pests.

In 1995, a GM maize (corn) was approved in the USA using the *Bacillus thuringiensis* gene to express a toxin to kill insects feeding on the plants. This provided a very efficient way of delivering the insecticide as it can kill first instar larvae of lepidopteran pests as soon as they start to eat the GM plants. This was soon followed by Bt cotton and Bt soybeans that were thus protected from important pests without the need to spray the crop. Unfortunately, in some situations farmers were not adequately prepared for sucking pests that did require a separate insecticide treatment, either as a spray or a seed treatment to protect young plants. Bt cotton in India, for example, was sprayed to control aphids and jassids using broad-spectrum insecticides that resulted in loss of natural enemies and infestations of insects previously not usually associated with cotton.

In deciding to adopt GM Bt cotton, a key factor must be whether the variety is suited to the environmental conditions where it is grown. Cotton growers in Burkina Faso were early adopters of GM cotton (Fig. 11.3) with seed obtained from Monsanto, but later it was realized the quality of the cotton was poor due to shorter fibres, emphasizing the importance of local research to select varieties. Monsanto was sued for US$83.91 million and growers recompensed by funds due to be paid as royalties for using the Bt seed, but this has put growers off growing GM cotton and they have reverted to older varieties selected for west African conditions.

Growing Bt crops is likely to result in key pests becoming resistant to the Bt toxins. In Australia, *Helicoverpa armigera* has developed resistance to nearly all the insecticides applied to the crop within five to eight years, so in introducing Bt cotton, initially with one insecticidal gene followed by Bollgard II in 2003 and Bollgard 3, with Cry1Ac, Cry2Ab and Vip3A genes, in 2016, they have adopted two programmes aimed at delaying the selection of resistant bollworms. The Genetic Dilution programme involves growing non-Bt crops, such as pigeon pea as a 'refuge' crop to produce sufficient susceptible genes to dilute resistant genes in bollworms from Bt cotton. Season quarantining in cooler areas where pupae may enter a diapause involved using trap crops that can be destroyed and the land ploughed to destroy pupae. Further studies on the attractiveness of non-Bt crops is needed but dilution of the resistant population with

Fig. 11.3. Field of genetically modified Bt cotton in Burkina Faso, 2008.

susceptible genes is clearly crucial in delaying resistant bollworms not being controlled (Whitehouse *et al.*, 2017).

Apart from the main type of *Gossypium hirsutum*, when growing *Gossypium barbadense*, grown mostly under irrigation, as in Gezira in Sudan, the naturally higher gossypol content made it less attractive to *Helicoverpa armigera* (bollworm), but when sprays were applied to control bollworms, heavy infestations of whiteflies occurred, resulting in 'sticky' cotton due to excessive honeydew on the open bolls. Perhaps a different approach to the genetic manipulation of cotton varieties would be more effective than using the Bt toxins. Undoubtedly, the gossypol content is a key protection of cotton plants, as glandless cotton was soon attacked by pests normally associated with maize.

Growing Bt crops soon caused concern among environmentalists as it was shown in laboratory tests that the *Bacillus thuringiensis* toxin can kill butterflies. However, butterfly larvae tend to feed on specific plants, not necessarily the crops engineered to contain Bt toxin, so they would not be exposed in the same way as crop pests. Follow-up field studies in farming areas subsequently confirmed the safety of the technology.

Alongside Bt crops, Monsanto developed the first crop to be genetically engineered to tolerate a specific herbicide. Soybean tolerant to the herbicide glyphosate was introduced in 1996, which made weed management much easier for farmers as they could delay a treatment until after the crop was established. Maize, cotton and other glyphosate-tolerant crops soon followed (Fig. 11.4) which inevitably led to gross overuse of the same herbicide, Roundup, and weeds becoming resistant to it (Fig. 11.5). The manufacturers of agrochemicals soon looked to the development of new

Fig. 11.4. GM maize and soybeans in rotation. (Photo courtesy of Santiago del Solar Dorrego, used with permission).

Fig. 11.5. Super weeds in soybean crops in the USA. (Photo courtesy of Bob Hartzler, Iowa State University).

varieties of GM crops that would tolerate alternative herbicides. Thus there was renewed interest in 2,4-D and dicamba, although it was recognized that these needed a different type of formulation to avoid the risk of volatile components causing damage to susceptible crops downwind. Less suitable, and no doubt less expensive, formulations of dicamba used by some farmers soon resulted in claims for damage to neighbouring crops, leading to further restrictions of its use.

The initial development of GM crops was due to the commercial investment in a new technology and it was, not surprisingly, perceived initially by the public that the only ones to benefit would be the agrochemical companies. The technology can provide other improved attributes of much more interest to the public looking for increased nutritional value, and to farmers wanting drought tolerance and higher yields. One early development in 2000 was 'golden rice', which increased vitamin A in rice where it is a dominant part of people's diet. Deficiency of vitamin A was considered to kill over 500,000 people every year, especially in areas of Asia. It could be argued that vitamin A could be given as a tablet, but mass drug distribution does not necessarily reach all those who need it, especially in the more remote areas, whereas enhanced vitamin A in GM crops enables people to get it as part of their everyday diet. Other crops being developed to enhance vitamin A include bananas and chickpea. Chinese scientists have also developed a purple rice with high levels of antioxidants.

Whatever improvements to diet or drought tolerance can be achieved, a reduction in pesticide use and a simpler programme would be of enormous benefit to farmers. Disease resistance to minimize a need for fungicides is one area where traditional plant breeding has made a significant contribution already and should continue to do so. Tomatoes with resistance to powdery mildew and the small brown willow moth have been developed, while a new cassava variety has resistance to mosaic virus and brown streak disease. With a wide range of pathogens attacking crops, it may still be necessary to treat a GM crop tolerant to a major disease with a fungicide, if a seasonal factor causes an outbreak of a different pathogen.

According to Parisi *et al.* (2016), there were 102 GM events authorized in at least one country, of which 49 were in commercial cultivation and 53 at the pre-commercial stage. Development of another 43 GM events had reached the regulatory stage and at least a further 77 were at an advanced research stage. The advent of Bt crops has undoubtedly allowed farmers to reduce their reliance on insecticide sprays, but not every pest is susceptible to the toxins currently used. In the future, development of alternative strategies to widen the activity of genetically incorporated toxins could be exploited to minimize the need for sprays. Further development of tolerance to specific herbicides should expand the ability of farmers to simplify their weed management programmes. In both these scenarios, more than one mode of action is essential to avoid selection of 'superweeds' as has occurred with gross overuse of glyphosate.

Detailed genome studies of plants may well reveal pathways that can be exploited by plant breeders whether using genetic transformation techniques or more traditional selection of new varieties to increase resistance to plant diseases and insect pests. Whatever technique is used, the varieties developed need to be grown within the context of integrated pest management as tolerance to one group of pests may result in susceptibility to other pests or affect other important characteristics such as the quality of the harvested crop.

Genetically Modified Insects

Knipling (1959) was the first person to advocate the release of modified insects to suppress natural populations of pest insects. In the USA, the release of sterile insects was highly successful in controlling the screwworm, *Cochliomyia hominovorax*, an obligatory parasite of livestock. Pupae were irradiated and after an initial trial on the island of Curacao, when 400 sterile males were released per square mile, the technique was used from Florida across the southern states into central America. The programme was very successful until 1972–1976 and again in 1978, due to genetic diversity of the screwworms, but changes enabled successful control, which has continued since 1979 (Richardson *et al.*, 1982), and the technique was also used in north Africa to suppress an outbreak of the insect to prevent its spread to other parts of Africa (Vargas-Teran *et al.*,

1994). The technique has been used for other insects, including fruit flies, tsetse flies and pink bollworm.

However, there is now more interest in release of genetically modified insects. The vector of dengue disease, *Aedes aegypti*, has been modified by inserting an inherited dominant lethality trait that results in the death of the larvae. The release of insects with a dominant lethal gene (RIDL) involves breeding large numbers of mosquitoes, providing tetracycline to larvae so they survive, but when the adults are released in the field their offspring do not have access to the tetracycline, and die. Large numbers of the modified mosquitoes need to be released to compete with the wild population that declines as the larvae fail to survive. The technique is self-limiting, thus it is necessary to repeat releases to suppress a target population (Alphey and Alphey, 2014). It may be necessary to reduce the natural population in an area prior to release by using an area-wide non-residual space spray insecticide treatment to enable the release of fewer GM mosquitoes.

An alternative procedure is to release mosquitoes in which a self-sustaining system is used; a genetic element is expected or designed to persist indefinitely and perhaps increase in frequency, and possibly invade other populations or species. This approach is being explored with the aim of controlling the *Anopheles* mosquitoes that transmit malaria, where there may be several species occuring within a single village. At present, some trials in Australia have involved releases of *Aedes aegypti* with *wolbachia*, and other research is examining the use of homing endonuclease genes (HEG) (Alphey, 2014). (HEGs are selfish DNA elements encoding proteins (endonucleases) that recognize and cleave specific DNA sequences of ~20–30 nucleotides.)

Hammond *et al.* (2016) described CRISPR-Cas9 endonuclease constructs that function as gene drive systems in *Anopheles gambiae* and identified three genes that confer a recessive female sterility. Population modelling and cage experiments have indicated that one of these genes meets the minimum requirement for a gene drive that targets female reproduction and which could suppress mosquito populations to levels so that malaria transmission is not supported. Alongside the research, there is attention on how these new approaches should be regulated.

Vertical Farming

With the Green Revolution helping farmers with fertilizers and pesticides, there has been an impressive increase in global grain production since the 1960s, but 795 million people are 'food-insecure' and an estimated 2 billion people are prone to malnutrition. The global population is expected to rise from 7.4 billion in 2016 to 9.7 billion by 2050. This increase is occurring mostly in the developing countries. The FAO and others are saying that global food production needs to increase by 60–70% between 2005 and 2050.

In addition to increasing yields by improved sustainable agriculture on existing croplands, according to Lal (2016), the strategy has to simultaneously reduce food waste, increase access and distribution of food and promote a plant-based diet. The aim is to reconcile high production with better environmental quality, and also develop urban agriculture with aquaponics, aeroponics and vertical farms (Fig. 11.6). It is claimed that 'sustainable intensification' of agro-ecosystems can produce enough food grains to feed one person for a year on 0.045 ha of arable land.

The area of agricultural land available for farming is now restricted with recognition that forests need to be conserved, and many areas of land have been degraded by soil erosion, encroachment of desert and salinity of ground water. This justifies the idea of producing some crops within

Fig. 11.6. Vertical farming of lettuce. (Photo by Valcenteu. Reproduced under Creative Commons Attribution-Share Alike 3.0 Unported license: https://commons.wikimedia.org/wiki/File:VertiCrop.jpg.)

cities under controlled lighting systems, using hydroponics and aeroponics, often in places with a limited ground space, but with the potential for expanding vertically. This is not necessarily a new concept as agriculture has been integrated into cities for many centuries. With the floating gardens in Mexico, the Aztecs converted the marshy wetlands of Lake Texcoco into chinampas, which were artificial islands created by building up the soil in the lake, to feed their growing population. This represents a masterpiece of engineering. Farming has always interacted with cities, as farmers had urban markets to supply and could use human waste as manure (Lawson, 2016). Modern hydroponic vertical farming began in Japan and city states like Singapore.

Urban horticulture has increased globally with an estimated 100 million people involved in growing vegetable crops with potential yields up to 50 kg/sq.m/year (Eigenbrod and Gruda, 2015). Much of this is growing small areas of land, sometimes referred to as allotments, allocated to individuals or used communally (Mok *et al.*, 2014), being unsuitable for housing, but always under threat from local development. A major concern is that these areas may be on land polluted with industrial chemicals and known as 'brownfield' sites, using polluted water, and that the growers may be applying pesticides with little or no knowledge about their correct use. Sampling vegetable crops (lettuce, cabbage and spring onion) in Ghana, in a peri-urban area, revealed a pesticide residue on lettuce leaves as well as high total and fecal coliforms and helminth egg counts on all three vegetables (Amoah *et al.*, 2006). Some countries, such as Cuba, have promoted 'organic' in peri-urban areas to maintain supplies when other sources had failed. The advantage of indoor vertical farming is that nutrients can be supplied more accurately and that use of pesticides can be avoided. Nevertheless, indoor farming is expensive and should not be regarded as 'local produce' at any price, but it will continue to be important to sustain traditional supplies of vegetables. Outdoor peri-urban farming will require support with genetically developed crops resistant to pests to avoid misuse of pesticides within this environment.

The Forward Path

The world and the farming scene will continue to change, but in looking ahead it is useful to look back, too. In the 1940s, with World War II, every effort was made to increase food production in the UK. Unused land and grassland was ploughed up to grow a larger area of cereals and other crops. The government urged everyone to 'dig for victory'. Yields were low as the improved varieties and pesticides were still only on the horizon, so wheat yields were only about 2 tons/ha. Birds and other wildlife, such as rabbits, thrived on the vast area of low-intensity farmland with plenty of insect pests and weeds. In the UK, following the setting up of several research institutes at the time of World War I, the Agricultural Research Council was formed, in 1931, which later became the Agricultural and Food Research Council in 1983. The Council, from the mid-1940s for two to three decades, supported practical field research, which undoubtedly played a crucial part in integrating pesticides into the farming system, but once there seemed to be a surplus of some commodities, the government closed most of the internationally recognized research organizations. Their research had enabled farmers today to harvest yields that are five to eight times higher, thanks to our ability to control weeds and other pests. For environmentalists, the sad story is that arable fields are no longer the wildlife-rich environment they once were. It is the balance between the needs of man and his environment that requires that crop yields in the future are achieved with more scientific research to refine the technology. There is now a gap in the supply of sufficient independent crop protection research to sort out how we can maintain high yields, whether by genetic engineering or other avenues, and an increasing reliance on commercial enterprises at a time when a sector of the public is highly suspicious of what industry recommends. Where pesticides continue to be needed, systems are required with targeted application, improved timing of sprays and greater use of geographical zones in which pesticide use is controlled to conserve existing pesticides, rather than continuing to ban important pesticides without detailed assessment of the consequences and losing, potentially, an important tool in the control of pests.

Despite the prediction of a 'silent spring', changes in pesticides and the registration of them over the last five decades have cut out the most hazardous and persistent in the environment; so birds continue to sing.

References

Alphey, L. (2014) Genetic control of mosquitoes. *Annual Review of Entomology* 59, 205–224.

Alphey, L. and Alphey, N. (2014) Five things to know about genetically modified (GM) insects for vector control. *PLOS Pathogens* 10(3), e1003909.

Amoah, P., Drechsel, P., Abaidoo, R.C. and Ntow, W.J. (2006) Pesticide and pathogen contamination of vegetables in Ghana's urban markets. *Archives of Environmental Contamination and Toxicology* 50, 1–6.

Belmain, S.R., Amoah, B.A., Nyirenda, S.P., Kamanula, J.F. and Stevenson, P.C. (2012) Highly variable insect control efficacy of *Tephrosia vogelii* chemotypes. *Journal of Agricultural and Food Chemistry* 60, 10055–10063.

Bilgo, E., Lovett, B., Fang, W., Bende, N., King, G.F., Diabate, A. and St Leger, R.J. (2017) Improved efficacy of an arthropod toxin expressing fungus against insecticide resistant malaria-vector mosquitoes. *Scientific Reports* 7, 3433.

Davies, S.-A. (2017) nEUROSTRESSPEP: novel biocontrol agents for insect pests from neuroendocrinology. *International Pest Control* 59, 164–165.

Eigenbrod, C. and Gruda, N. (2015) Urban vegetable for food security in cities: a review. *Agronomy for Sustainable Development* 35, 483–498.

Hammond, A., Galizi, R., Kyrou, K., Simoni, A., Siniscalchi, C. *et al.* (2016) A CRISPR-Cas9 gene drive system targeting female reproduction in the malaria mosquito vector *Anopheles gambiae*. *Nature Biotechnology* 34, 78–83.

He, X.K., Bonds, J., Herbst, A. and Langenakens, J. (2017) Recent development of unmanned aerial vehicle for plant protection in East Asia. *International Journal of Agricultural and Biological Engineering* 10(3), 18–30.

King, G.F. and Hardy, M.C. (2012) Spider-venom peptides: structure, pharmacology, and potential for control of insect pests. *Annual Review of Entomology* 58, 475–496.

Knipling, E.F. (1959) Sterile-male method of population control. *Science* 130, 902–904.

Lal, R. (2016) Feeding 11 billion on 0.5 million hectare of area under cereal crops. *Food and Energy Security* 5, 239–251.

Lan, Y., Shengde, C. and Fritz, B.K. (2017) Current status and future trends of precision agricultural aviation technologies. *International Journal of Agricultural and Biological Engineering* 10(3), 1–17.

Lawson, L. (2016) Sowing the city. *Nature* 540, 522–524.

Mok, H.-F., Williamson, V.G., Grove, J.R., Burry, K., Barker, S.F. and Hamilton, A.J. (2014) Strawberry fields forever? Urban agriculture in developed countries: a review. *Agronomy Sustainable Development* 34, 21–43.

Nakasu, E.Y.T., Williamson, S.M., Edwards, M.G., Fitches, E.C., Gatehouse, J.A., Wright, G.A. and Gatehouse, A.M.R. (2014) Novel biopesticide based on a spider venom peptide shows no adverse effects on honeybees. *Proceedings of the Royal Society*, Series B 281, 20140619.

Parisi, C., Tillie, P. and Rodríguez-Cerezo, E. (2016) The global pipeline of GM crops out to 2020. *Nature Biotechnology* 34, 31–36.

Richardson, R.H., Ellison, J.R. and Averhoff, W.W. (1982) Autocidal control of screwworms in North America. *Science* 215, 361–370.

Ullah, I., Hagenbucher, S., Alvarez-Alfageme, F., Ashfaq, M. and Romeis, J. (2015) Target and non-target effects of a spider venom produced in transgenic cotton and tobacco plants. *Journal of Applied Entomology* 139, 321–332.

Vargas-Teran, M., Hursey, B.S. and Cunningham, E.P. (1994) Eradication of the screwworm from Libya using the sterile insect technique. *Parasitology Today* 10, 119–122.

Wang, C. and St Leger, R.J. (2007) A scorpion neurotoxin increases the potency of a fungal insecticide. *Nature Biotechnology* 25, 1455–1456. DOI: 10.1038/nbt1357.

Whitehouse, M.E.A., Cross, D., Mansfield, S., Harden, S., Johnson, A.L., Harris, D.J., Tan, W. and Downes, S.J. (2017) Do pigeon pea refuges in *Bt* cotton pull their weight as 'genetic diluters' to counter *Bt* resistance in *Helicoverpa* moths? *Crop Protection* 100, 96–105.

Windley, M.J., Herzig, V., Dziemborowicz, S.A., Hardy, M.C., King, G.F. and Nicolson, G.M. (2012) Spider-venom peptides as biopesticides. *Toxins* 4, 191–227.

Annex

Common Name and Major Trade Name of Selected Pesticides

The WHO Classification is based on the list published by WHO in 2009.

Common Name	Type	Trade Name	WHO Classification
abamectin	I A	Agrimec	
aldicarb	I N	Temik	Ia
aldrin	I	Aldrex	Now obsolete
atrazine	H	Gesaprim	III
azadirachtin (neem)	I		
azinphos methyl	I	Guthion	Ib
azoxystrobin	F	Amistar	U
Bacillus thuringiensis	I		III
benomyl	F	Benlate	U
bifenthrin	I	Talstar	II
Bordeaux mixture	F	Bordocop	
captafol	F	Difoltan	Ia
captan	F	Captaf	U
carbaryl	I	Sevin	II
carbofuran	I	Furadan	Ib
chlorantraniliprole	I	Acelepryn	U
chlorfenapyr	I	Pirate	II
chlorpyrifos	I	Dursban	II
clothianidin	I	Deter	
copper sulfate	F	Blue Viking	II
cyfluthrin	I	Baythroid	Ib
dalapon	H	Dowpon	U
2,4-D	H		II
DDT	I	DDT	II

Continued

Common Name	Type	Trade Name	WHO Classification
deltamethrin	I	Decis	II
demeton-S-methyl	I	Metasystox	Ib
dicamba	H	Banvel	II
dichlorvos	I	DDVP; Vapona	Ib
dieldrin	I		Now Obsolete
dimethoate	I	Rogor	II
endosulfan	I	Thiodan	II
endrin	I	Endrex	Now obsolete
fenitrothion	I	Sumithion	II
fenthion	I	Lebaycid	II
fipronil	I	Regent	II
gamma BHC	I	Lindane	II
glufosinate	H	Liberty	II
glyphosate	H	Roundup	III
imidacloprid	I	Confidor	II
isoproturon	H	Tolkan	II
lambda cyhalothrin	I	Karate	II
lead arsenate	I		Ib
Lufenuron	I	Match	III
malathion	I	Fyfanon	III
mancozeb	F	Dithane	U
MCPA	H	Agroxone	II
methidathion	I	Supracide	Ib
methomyl	I	Lannate	Ib
methyl bromide	Fumigant	Brom-O-Gas	Not classified
methyl parathion	I	Folidol	Ia
monocrotophos	I	Nuvacron	Ib
nicotine	I		Ib
oxamyl	I	Vydate	Ib
paraquat	H	Gramoxone	II
permethrin	I	Ambush	II
phorate	I	Thimet	Ia
piperonyl butoxide			U
pirimicarb	I	Aphox	II
pirimiphos-methyl	I	Actellic	III
profenophos	I	Curacron	II
quinalphos	I	Danulux	II
rotenone	I		II
simazine	H	Gesatop	U
sodium cyanide			Ib
spinosad	I	Tracer	III
spirotetramat	H	Movento	III
tebuconazole	F	Folicur	II
teflubenzuron	I	Nomolt	III
tefluthrin	I	Force	Ib
temephos	I	Abate	III
thiodicarb	I	Larvin	II
thiram	F	Thiram	II

Continued

Common Name	Type	Trade Name	WHO Classification
triazophos	I	Hostathion	Ib
trichlorphon	I	Dipterex	II
trifluralin	H	Treflan	U

A= Acaricide; I = Insecticide; F = Fungicide; H = Herbicide

WHO Class for active ingredient. Ia – Extremely hazardous; Ib Highly hazardous; II – Moderately hazardous; III – Slightly hazardous; U – Unlikely to present acute hazard. Note the class is also affected by the amount used in a formulated product, thus the low dose of pyrethroids significantly changes the classification.

The trade name mentioned is usually that given by the company that did the original development, but there are often many different trade names for individual actives and when used in a mixture.

Index

Page numbers in **bold** type refer to figures, tables and boxed text.